Analysis of Species-Specific Molar Adaptations in Strepsirhine Primates

Contributions to Primatology

Vol. 11

Editor: F. S. SZALAY, New York, N.Y.
Associate Editors: P. CHARLES-DOMINIQUE, Brunoy; H. KUHN, Göttingen;
W. P. LUCKETT, Omaha, Nebr.
Founding Editors: A. H. SCHULTZ †, Zürich; D. STARCK, Frankfurt
am Main

S. Karger · Basel · München · Paris · London · New York · Sydney

Analysis of Species-Specific Molar Adaptations in Strepsirhine Primates

DANIEL SELIGSOHN, Brooklyn, N.Y.

68 figures and 9 tables, 1977

S. Karger · Basel · München · Paris · London · New York · Sydney

Contributions to Primatology

Manuscripts should be sent to: Dr. F.S. Szalay, Department of Anthropology, Hunter College, 695 Park Avenue, New York, NY 10021 (USA)

Vol. 8: CONROY, G.C. (New York, N.Y.): Primate Postcranial Remains from the Oligocene of Egypt. VI + 134 p., 38 fig., 36 tab., 1976.
ISBN 3-8055-2333-5
Vol. 9: FEDIGAN, L.M. (Edmonton, Alta.): A Study of Roles in the Arashiyama West Troop of Japanese Monkeys *(Macaca fuscata)*. VII + 96 p., 5 fig., 15 tab., 1976.
ISBN 3-8055-2334-3
Vol. 10: CRAMER, D.L. (New Brunswick, N.J.): Craniofacial Morphology of *Pan paniscus*. VI + 64 p., 21 fig., 8 tab., 1977.
ISBN 3-8055-2391-2

Cataloging in Publication
    Seligsohn, Daniel
    Analysis of species-specific molar adaptations in Strepsirhine primates
    Daniel Seligsohn. Basel, New York, Karger, 1977.
    (Contributions to primatology, v. 11)
    1. Molar 2. Prosimians I. Title II. Series
    W1 C0778UP v. 11/ QL 737.P9 S465a
    ISBN 3-8055-2634-2

Contents

Acknowledgments

Among those people whose help and encouragement enabled me to complete this study, I offer special thanks and gratitude to FREDERICK S. SZALAY for his invaluable criticism and advice, and for the enthusiasm he instilled within me for functional dental morphology.

I am also indebted to SIDNEY ANDERSON and MALCOLM C. McKENNA for allowing me to study the crania of the living and subfossil strepsirhines at the American Museum of Natural History.

To my dear wife, BELLA, I offer loving thanks for her unceasing patience, forebearance, devotion and moral support, and for the expert typing skill which she offered toward the completion of this work.

To my parents, my special thanks for being there through the best of times and through the worst of times.

I. Introduction

Mammalian molar teeth are highly calcified oral organs which demonstrate great morphological diversity. As recent cineradiographic studies have shown, these structures are in intimate contact with ingested food during chewing. An inference drawn from these and from occlusal studies is that upper and lower molars occlude with great precision, and that such precision is indicative of the importance of the functional efficiency of these structures. It has long been recognized that the primary biological role of molar teeth is to mechanically divide ingested food so as to facilitate chemical digestion in the gastrointestinal tract. That this biological role is a crucial prerequisite to the rapid digestion of tough foods is graphically demonstrated by the tendency of many diverse taxa to evolve mechanisms to compensate for the lack of functional mandibular teeth, i.e., pyloric 'teeth' in pangolins [KRAUSE and LEESON, 1974] and gizzard stones in some reptiles and birds.

Functional dental morphology has arisen as a distinct discipline only relatively recently, and seeks to functionally explain the relationship between molar form and occlusion and dietary preference within the context of the total mammalian feeding mechanism. Living forms have provided a laboratory within which to test hypotheses in this area. While such efforts have been significant in and of themselves, they have provided the raw material from which broader principles of molar function can be established. These principles can then be applied to the solution of problems in a most exciting area: the mammalian fossil record.

The vast bulk of the mammalian fossil record consists of (molar) teeth. This fact dictates that most attempts to explain the fossil record must resort to concepts developed within the field of functional dental morphology. So applied, these concepts provide precious evidence which bears not only on the dietary behavior of fossil taxa, but also on the functional and ecological significance of mammalian phylogeny.

As with any discipline which deals with evolutionary adaptation, functional dental morphology provides no ready method of recognizing biological role (dietary preference) once (molar) function is understood. A firm understanding of comparative dental morphology and of the systematics of fossil taxa are required if interpretations of biological role are to be soundly based.

Most early (and some recent) studies in functional dental morphology sought to establish and functionally explain molar occlusal relationships and to infer mandibular chewing movements from molar wear features [e.g., RYDER, 1878; COPE, 1889; OSBORN, 1911; GREGORY, 1920, 1922; SIMPSON, 1936; BUTLER, 1952, 1972, 1973; MILLS, 1955, 1966; GINGERICH, 1972]. Efforts were also made to reconstruct the evolution of mammalian molar morphology and molar occlusion [e.g., PATTERSON, 1956].

Major technological breakthroughs in research techniques have recently permitted *in vivo* cineradiographic and cinefluorographic analyses of mandibular movements during mastication in several disparate mammalian taxa [e.g., ARDRAN *et al.*, 1958; HIIEMAE and ARDRAN, 1968; CROMPTON and HIIEMAE, 1970; HIIEMAE and CROMPTON, 1971; KALLEN and GANS, 1972; HIIEMAE and KAY, 1972, 1973; KAY and HIIEMAE, 1974a, b]. These and other similar studies have furnished valuable data which have permitted a much clearer understanding of the jaw movements, molar wear and molar occlusion in mammals, and how they interrelate.

These studies have shown that most mammals demonstrate very similar patterns of jaw movements during feeding. A bite of food is usually initially acquired by various ingestive procedures (depending upon food consistency). If teeth are required, these procedures may involve the anterior or postcanine dentition. Food is then mechanically divided or deformed during mastication which involves the postcanine dentition and two distinct phases of repetitive and rhythmic jaw movements. Both phases involve a preparatory stroke which elevates the jaw from maximum gape and ultimately brings the teeth either into forceful contact with food or into occlusion. The power stroke then follows during which the teeth are brought into closest apposition or into occlusion and during which teeth apply maximum force to the food. The recovery stroke then follows, during which the teeth move out of functional apposition or occlusion and the jaw drops to maximum gape. Mastication is usually comprised of two successive phases: puncture-crushing and chewing. Both usually involve only one side of the jaw during any given interval. Puncture-crushing is characterized by a more orthal preparatory stroke and involves the initial division or deformation of food between premolars and molars which closely interdigitate but do not occlude. Unless food is very soft (in which case mastication ceases at this point), this phase then grades into chewing during which food is finely divided, primarily by the molar teeth. In mammals with molars relatively little modified from the primitive therian tribosphenic plan, chewing is most notably characterized by a greatly increased bucco-lingual component to the power stroke. The power stroke is divided into two successive phases: the buccal and the lingual. The buccal phase involves a dorso-linguo-mesial movement of the jaw during which the molar crest edges achieve occlusion and point cut (i.e., shear) food. The buccal phase terminates with centric occlusion during which interposed food is, loosely speaking, crushed. The lingual phase follows, characterized by a ventro-mesio-lingual jaw movement which brings the molars out of occlusion and, loosely speaking, permits food to be ground between molar basin surfaces. These studies generally assume: that abrasive wear especially on the cusp apices of molar teeth results largely from tooth-food-tooth contact during the puncture-crushing phase of mastication; that striated, facetted wear along the edges of molar crests (and hence, presumably edge sharpening) results from occlusal contacts of molar crests during buccal phase chewing; and that wear features on the surfaces of molar basins are due to lingual phase jaw movements. These studies also suggest that molar morphology is a major factor in determining mandibular movements during the power stroke of chewing.

The cinefluorographic and cineradiographic studies have permitted a greater functional appreciation of the evolution and occlusion of the therian tribosphenic molar and of the wider modifications in the mammalian feeding mechanism [e.g., CROMPTON and HIIEMAE, 1969; CROMPTON and SITA-LUMSDEN, 1970; CROMPTON, 1971].

The still unresolved problem of how mammalian molars maintain sharp cutting edges throughout ontogeny has been grappled with very creatively by EVERY [EVERY, 1960, 1970, 1972, 1974; EVERY and KUHNE, 1971] who has emphasized the great importance of point-cutting to chewing. EVERY has proposed that molar crests are sharpened by a jaw movement which is opposite in direction to that of the chewing power stroke, and temporally distinct from mastication. He has termed this phenomenon 'thegosis'. This investigator has also provided keen insight into the mechanics of molar occlusion.

Inspired by EVERY, GRAHAM [1969] conducted a study of the dentitions of a population of *Didelphis*. As part of her study, she described the pattern of ontogenetic wear in the molars of *Didelphis*, and demonstrated compensatory changes in the molar crests which helped maintain cutting efficiency despite ontogenetic wear.

Only very recently have groups of mammals been studied for the purpose of examining the functional relationship between molar morphology and the physical properties of preferred foods. WALKER and MURRAY [1975], for example, demonstrated that colobines more efficiently masticate leaves than do cercopithecines, and they attempted to relate this fact to the morphological differences in the molars of these two subfamilies.

In an effort to determine diet-related adaptations in the second molars of noncercopithecid primates, KAY [1973, 1975] employed multivariate statistics to compare species with respect to the size of several molar features, while controlling for body mass. KAY concluded that the molars of highly insectivorous and highly folivorous primates demonstrate equivalent cutting, crushing and grinding capabilities, and that they cannot be distinguished on the basis of their molar features alone. KAY had to utilize the average difference in body mass between insectivorous and folivorous primates to separate these two dietary groups. KAY also concluded that highly frugivorous primates possess smaller molars, and hence lower overall capabilities of molar function, than do highly insectivorous or highly folivorous primates. In a later study, KAY [in press] clarified the pattern of molar occlusion in cercopithecid primates and concluded that for equivalent molar lengths, the second molars of more folivorous cercopithecids demonstrate greater cutting, crushing, and grinding capabilities than do the molars of more frugivorous taxa. He also found that the cercopithecines (being more frugivorous) unexpectedly possess larger molars relative to body mass than do the colobines (being more folivorous). KAY in this study was unable to establish a dietary separation of cercopithecid molars by relating the size of their molar features to body mass. Clearly, KAY's approach cannot be applied to isolated fossil teeth.

The present study attempts to add to the understanding of the functional relationship between molar morphology and dietary preference. In essence, this study involves the morphological and metrical comparison of the upper and lower second molars of most strepsirhine species, and centers especially on those species for which adequate dietary data are available. This study is guided by several models of molar adaptation. The interspecific comparison of molars provides the means of testing these models as well as of delineating the relative influences of molar size, heritage and dietary preference on molar form. The major thrust of this study is to explore the relationship between the functional morphology of molar teeth and the physical properties of preferred foods. Some attention is also given to ontogenetic patterns of molar wear and molar reorientation. This study, it is hoped, will have maximum applications to the problem of inferring dietary preference from fossil molar teeth, as this study does not depend on body mass.

The M^2 and $M_{\overline{2}}$ were chosen as the objects of this study because these teeth are situated in the middle of the molar regions of the upper and lower jaws, where food is most finely chewed. This would appear to indicate that the second molars most directly reflect selective demands on chewing efficiency. The strepsirhine primates were chosen for this study because, while sharing a common tooth-combed ancestor (i.e., they are a monophyletic group), they demonstrate a great diversity in molar morphology and dietary behavior. This situation is ideal to the investigation of diet-related molar adaptations.

II. Hypotheses

A. Assumptions

My hypotheses of strepsirhine molar adaptations seek to functionally relate molar form to dietary preference. These hypotheses were formulated within a theoretical framework which views molar teeth as form-function complexes [*sensu* BOCK and VON WAHLERT, 1965]. Several assumptions have been made in framing my hypotheses: (1) that the primary biological role of molar teeth is to increase the surface area of ingested food by mechanically dividing it, thus assisting chemical digestion; (2) that within the context of the given morphogenetic machinery (or heritage) of a species, natural selection will favor the most (energy) efficient interaction between a feature of an organism and a feature of the environment; (3) that within the constraints imposed by heritage, the physical properties of ingested food constitute significant selective pressures on molar form; that insects, fruits, leaves and stems possess the physical properties I have ascribed to them in my hypotheses; and that the dietary preferences now observed in the wild apply to the geological past; (4) that molar teeth are evolutionary compromises in the sense that their evolutionary adaptations result from interactions between many (sometimes conflicting) selection pressures and heritage features (molar teeth, then, are finite, evolutionarily integrated structures, and have definite limits set on their morphogenetic and physiological capabilities); (5) that the differences in morphology of homologous molar features at least partly reflect differences in the capabilities of these features to mechanically divide masticated food; (6) that mandibular kinematics, molar occlusion and the relationship between molar wear and molar occlusion within the strepsirhines are as demonstrated (or inferred) in mammals thus far studied in these respects [e.g., CROMPTON and HIIEMAE, 1969; CROMPTON and SITA-LUMSDEN, 1970; KALLEN and GANS, 1972; HIIEMAE and KAY, 1973; KAY and HIIEMAE, 1974b]; (7) that the currently accepted phylogeny of the strepsirhines [SZALAY and KATZ, 1973; SZALAY, 1975; CARTMILL, 1975] is accurate.

In addition to the above assumptions, several definitions and concepts relating to the dental mechanical division of food were invoked in framing my hypotheses. Many of these ideas have evolved within the field of functional dental morphology, beginning with the works of GREGORY [1920, 1922] and continuing through the works of BUTLER [1952, 1972, 1973], CROMPTON and HIIEMAE [1969], EVERY [1970, 1972, 1974], RENSBERGER [1973, 1975], etc.

In reality, the mastication of food involves a multitude of simultaneous processes which are inextricably interrelated both functionally and evolutionarily. I have arbitrarily established two major areas of mechanical molar function within which I will concentrate:

1. Mechanical Division of Food

Mechanical division of food is here defined as the apposition or contact of, or movement between two occluding molar features resulting in the deformation or division of interposed food material.

2. Food Escapement

Food escapement is here defined as the unimpeded movement of deformed or divided food material normal to the force vector of the occluding molar features performing the division of food. This movement of food occurs away from the occluding molar features.

Food escapement can be viewed as a mechanism which enhances the division of food. Increased food escapement permits a greater increase in the surface area of divided (or deformed) food, while it reduces the level of food resistance to the chewing force. Patterns of food escapement reflect both the physical properties of preferred foods as well as the modes of mechanical division employed to increase the surface area of these foods.

3. Types of Food Division and Escapement

Food division and escapement can be subdivided into several ideal models of molar function. (When referring to either the mechanical division of food or to food escapement, the term 'vertical' signifies a relationship roughly perpendicular to the cervical plane of the molar, while the terms 'horizontal' and 'lateral' signify a relationship roughly parallel with the cervical plane of the molar.)

a) 'Vertical' Point-Cutting

'Vertical' point cutting (fig. 1A) is a means of dividing (thick, tough) food at very high pressures by interposing food between two occluding sharp-

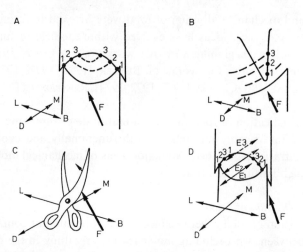

Fig. 1. 'Vertical' point-cutting and attendant pattern of food escapement. *A* and *B* schematically illustrate 'vertical' point-cutting with the points numbered 1–3 indicating the points at which the occluding crests cross (and point-cut food) at three successive instants in time. The arrows marked F indicate the movement of the lower molar feature relative to the upper molar feature. *A* schematically illustrates the nature of reciprocal edge curvature associated with 'vertical' point-cutting, while *B* schematically illustrates the nature of differential crest orientation associated with this form of food division. *C* presents as a rough analogy a pair of scissors oriented in a manner similar to molar crests performing 'vertical' point-cutting. The arrow marked F indicates the path along which the force acts to close of the pair of scissors. *D* schematically illustrates the pattern of food escapement associated with vertical point cutting. Arrows labelled E_1–E_3 indicate the axes along which food escapes during three successive instants in time as 'vertical' point-cutting progresses. The arrow marked F indicates the direction of the chewing force.

edged crests whose orientations strongly emphasize a component perpendicular to the cervical plane of the molar, and which are reciprocally curved and/or differentially oriented primarily in a plane which is both normal to that of the crown cervix and parallel with the bucco-lingual axis of the molar. Division of food involves the movement of these crests past each other. The space between the two sets of crests diminishes as these crests cross (at any given instant) at only one or two points. Point-cutting is thus capable of progressing relatively extensively along a vertical axis, while still utilizing the mandibular movements characteristic of primitive therian chewing.

The increase in surface area of food is proportional to the area encompassed by the two occluding crests and to the cross-sectional area of the food-material.

Food escapement during 'vertical' point-cutting (fig. 1B) occurs mainly laterally (parallel to the cervical plane) over a relatively great 'vertical' distance (i.e., normal to cervical plane of molar). This pattern of food escapement reduces food resistance to a minimum.

b) 'Horizontal' Point-Cutting

'Horizontal' point-cutting (fig. 2A) is a means of dividing (thin, tough) food at very high pressures by interposing food between two occluding sharp-edged crests whose orientations strongly emphasize both horizontal and mesio-distal components and which are reciprocally curved and/or differentially oriented primarily in a plane parallel with that of the crown cervix. Division of food involves the movement of these crests past each other. The space between the two sets of crests diminishes as these crests

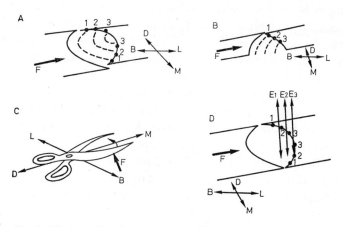

Fig. 2. 'Horizontal' point-cutting and attendant pattern of food escapement. *A* and *B* schematically illustrate 'horizontal' point-cutting, with points numbered 1–3 indicating the points at which the occluding crests cross (and point-cut food) at three successive instants in time. The arrows marked F indicate the movement of the lower molar feature relative to the upper molar feature. *A* schematically illustrates the nature of reciprocal edge curvature associated with 'horizontal' point-cutting, while *B* schematically illustrates the nature of differential crest orientation associated with this form of food division. *C* presents as a rough analogy a pair of scissors oriented in a manner similar to molar crests performing 'horizontal' point-cutting. The arrow marked F indicates the path along which the force acts to close the pair of scissors. *D* schematically illustrates the pattern of food escapement associated with 'horizontal' point cutting. Arrows labelled E_1–E_3 indicate the axes along which food escapes during three successive instants in time as 'horizontal' point-cutting progresses. The arrow marked F indicates the direction of the chewing force.

cross at only one or two points (at any given instant). Point cutting is thus capable of progressing relatively extensively along a horizontal (mesio-distal) axis, while still utilizing the mandibular movements characteristic of primitive therian chewing.

The increase in surface area of food is proportional to the area encompassed by the two occluding crests and to the cross-sectional area of the food material.

Food escapement during 'horizontal' point-cutting (fig. 2B) occurs mainly 'vertically' (normal to cervical plane) over a relatively great 'lateral' distance (i.e., parallel to cervical plane of molar). This pattern of food escapement reduces food resistance to a minimum.

c) Point Penetration

Point penetration (fig. 3A) is a means of puncturing (thick, tough) food at high pressures by interposing food between a pointed, cone-shaped feature and a basin-shaped feature. As the former feature normally approaches the latter one, the cone-shaped feature penetrates and 'wedges' apart the food material, while the basin-shaped structure confines the food to facilitate the mechanical impact of food penetration.

This process does not result in a great increase in the surface area of food, though the resulting increase is proportional to the height and width of the cone-shaped feature and to the cross-sectional area of the food, and is inversely proportional to the confinement of the basin-like structure. This process further facilitates the division of food material.

During point penetration food escapement (fig. 3B) laterally, normal to the chewing force, is restricted, thus resulting in escapement oblique or oppo-

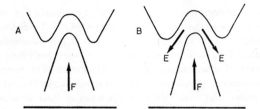

Fig. 3. Point penetration and attendant pattern of food escapement. A schematically illustrates the occlusion of molar features (seen in section) during point penetration. The arrow marked F indicates the movement of the upper and lower molar features relative to each other. The horizontal line below the drawing represents the cervical plane of the molar. B schematically shows food escapement during point penetration and uses the same conventions as A. The arrows marked E indicate the paths of food escapement.

site to the chewing force. Increased food confinement is achieved at the expense of increased food resistance, though levels of food resistance can be minimized if the cone-shaped feature is very tapered and pointed.

d) Crushing

Crushing (fig. 4A) is a very low pressure means of dividing (or deforming) food by interposing food material between two broad, planar surfaces, oriented as normal as possible to the chewing force. Food division (or deformation) involves the forceful approximation of the two planar surfaces, thus transmitting the greatest possible amount of chewing force to the greatest possible area of interposed food.

The increase in surface area of food is proportional to the area of the planar crushing surfaces and to the cross-sectional area of food material.

Food escapement during crushing (fig. 4B) is very extensive laterally (normal to chewing force). This pattern of food escapement minimizes the level of food resistance.

e) Grinding

Grinding (fig. 5A) is a very low pressure means of dividing (or deforming) food by interposing food material between two broad, planar surfaces which are oriented as normal as possible to the chewing force. Food division occurs as the two planar surfaces are forcefully and closely approximated while one surface moves parallel to the other. This arrangement transmits

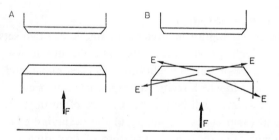

Fig. 4. Crushing and attendant pattern of food escapement. *A* schematically shows the occlusion of molar features characteristic of crushing. The arrow marked F indicates the movement of the lower molar feature relative to that of the upper molar. The horizontal line below the drawing represents the cervical plane of the molar. In *B* the same conventions as in *A* are used. *B* schematically illustrates the paths of food escapement during crushing which are indicated by arrows marked E. The chewing force is represented by the arrows marked F.

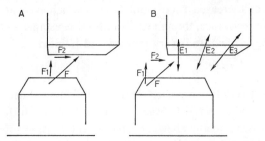

Fig. 5. Grinding and attendant pattern of food escapement. *A* schematically shows occluding molar features during grinding. The arrow labelled F indicates the actual movement of the lower molar feature relative to that of the upper molar. F_1 and F_2 are components of F, respectively normal and parallel to the occluding surfaces. The horizontal line below the drawing represents the cervical plane of the occluding molars. *B* schematically illustrates the pattern of food escapement during grinding. The conventions in *A* are used in *B*. In addition, arrows marking E_1–E_3 represent the paths of food escapement at three successive instants as grinding progresses.

the greatest possible amount of chewing force to the greatest possible area of interposed food, over the longest possible period of time.

The increase in the surface area of food is proportional to the area of the planar grinding surfaces and to the cross-sectional area of food material.

Food escapement during grinding (fig. 5B) is very extensive laterally (normal to chewing force) over a relatively great lateral distance. This pattern of food escapement minimizes the level of food resistance.

4. Some Comments on Point-Cutting

Point-cutting to be effective must involve the interaction of two sets of occluding cutting edges which demonstrate reciprocal curvature in two planes. The greater the total reciprocal edge curvature in each plane, and the greater the number of planes in which reciprocal edge curvature occurs (i.e., in one plane or two), the greater will be (1) the pressure of cutting, (2) the escapement of food, and (3) perhaps the ability of the occluding edges to hone themselves. In reality the reciprocal curvature of occluding cutting edges is usually better emphasized in one plane, and less emphasized in a second plane. Because two occluding edges are reciprocally curved in two given planes, they must move relative to each other in yet a third plane, if they are to occlude effectively during point-cutting (fig. 6A). This explains why buccal phase facetting (i.e., wear evidence of point-cutting) is always at an angle or bevel to the leading and trailing slopes of a crest.

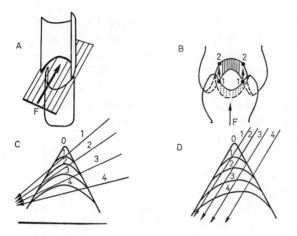

Fig. 6. Aspects of point-cutting and the maintenance of sharp molar cutting edges. *A* schematically illustrates the occlusion of molar crests which are reciprocally curved in two planes. The hatched rectangle represents the plane within which the lower molar crests move relatively to the crests of the upper molar. The arrow labelled F indicates the path along which the lower molar feature moves relative to that of the upper molar during buccal phase chewing. *B* illustrates an idealized model of point-cutting and edge-sharpening in two occluding molars. Hatched areas along the upper molar crest represent (buccal phase) facetting produced during the point-cutting of food which are on the molar surfaces facing the viewer. Interrupted lines indicate the same sort of facetting along the lower molar crest on the molar surface opposite to that visible. The points labelled 1 indicate the points along the occluding molar cutting edges at which point-cutting actually occurs at a given instant while the intervals marked $1 \rightarrow 2$ represent the distances over which these points on the cutting edges are sharpened or honed across the occluding molar facet subsequent to point-cutting. The arrow marked F indicates the movement of the lower molar relative to the upper molar. *C* schematically shows the pattern of honing of molar cutting edges which involves an ontogenetic increase in the horizontal component of buccal phase facetting. The triangular feature represents a bucco-lingual section at an upper molar crest, with lines labelled 0–4, representing crest surfaces between the time the crest is unworn to the time the crest is heavily worn. The arrows labelled 1–4 represent successively more horizontal axes along which the occluding lower molar crest moves relative to the crest of the upper molar. Arrows 1–4 correspond in time to surfaces 1–4 representing successive stages of wear in the upper molar crest. Arrows 1–4 thus indicate the orientation of buccal phase facetting which bevels or hones the cutting edges of the upper molar crest at four successive stages of crest wear. The horizontal line below the drawing represents the cervical plane of the molar. This mode of cutting edge sharpening results in the lowest possib.e loss of dental material. *D* schematically shows the pattern of honing of molar cutting edges which does not involve an ontogenetic increase in the horizontal component of buccal phase facetting. All conventions are as in *C*. *D* differs from *C* in illustrating edge honing with an ontogenetically constant orientation of buccal phase facetting. This mode of cutting edge sharpening results in an extravagant loss of dental material.

Edge sharpening (or 'thegosis') may be explained by the very nature of cutting edges which are reciprocally curved in two planes (fig. 6 B). After a point on a cutting edge cuts (i.e. 'coincides' in space and time with its occluding point), it is possible that it is honed when it contacts the occluding facet (basal to the occluding edge) during the 'free ride' subsequent to the actual cutting action. An increase in total reciprocal edge curvature would increase food escapement from between two cutting edges, decrease the duration and distance over which a given point actually cuts food, and (up to a point) increase the relative distance over which a point on an edge 'rides' against and is honed by an occluding wear facet. (Enamel chipped off a cutting edge may be trapped and utilized as an abrasive for honing.) A compromise may exist between selection pressures controlling the distance over which a point cuts and the distance over which a point is honed by an occluding facet.

The increased horizontal component of buccal phase facetting that often attends increased molar wear in mammals is a means of honing rounded and abraded crest edges by bevelling them (fig. 6 C). This results in the restoration of a sharp leading cutting edge without an unnecessary loss of dental material (fig. 6 D).

B. Models of Molar Adaptation

Tables I–V are five models which functionally relate molar morphology to dietary preference.

When referring to either molar form or molar function, the term 'vertical' signifies a relationship roughly perpendicular to the cervical plane of the molar, while the terms 'horizontal' and 'lateral' signify a relationship roughly parallel with the cervical plane of the molar.

All models, except that for 'crosslophed leaf-feeder', refer to molars which adhere relatively closely to the primitive mammalian tribosphenic plan.

The model for the 'crosslophed leaf-feeder' refers to a molar which is characterized by the following features: (a) possesses four subequal cusps which are organized into transversely aligned mesial and distal pairs; (b) the cusps within each pair are joined by a transverse ridge; (c) the morphology and relief of the buccal and lingual cusps in the M^2 and $M_{\bar{3}}$ are reciprocal.

Table I. Model of the molar adaptations of an insect- (and small vertebrate-) feeder[1].

Organism		biological role	↔	Environment	
molar characters	character states (chewing adaptations)			physical properties preferred foods	preferred foods
Cusps	*form* – very high relief, pointed and conical *function* – permit high pressure means of puncturing (and immobilizing) thick, tough food	assist in chemical digestion by producing greatest possible increase in surface area of ingested food	natural selection – favors most efficient interaction between organism and environment – given heritage	compact, thick, 3-dimensional, very tough, fibrous, variably compliant	insects and small vertebrates (accounting for nearly 100% of diet by weight per year)
Crests	*form* – their orientations strongly emphasize a component perpendicular to the cervical plane of the molar, while occluding crests are reciprocally curved and/or differentially oriented primarily in a plane which is both normal to that of the crown cervix and parallel with the bucco-lingual axis of the molar *function* – permit high pressure 'vertical' point-cutting to be carried out through relatively great thickness of thick, tough, food. Help reduce food resistance during point-cutting by facilitating horizontal food escapement over relatively great vertical distance, normal to chewing force				
Basins	*form* – deep, mesio-distally narrowed and confined with reciprocally concave surfaces mostly oriented oblique to chewing force *function* – facilitate mechanical impact of point penetration of thick, tough, food by restricting lateral food escapement. Reduce possible food resistance during point-cutting of food by permitting horizontal food escapement over a relatively great vertical distance, normal to chewing force. Efficient crushing and grinding are precluded				

[1] This model assumes a diet which is very restrictive. Further explanations are given in 'Hypotheses'.

Table II. Model of the molar adaptations of a fruit- (and gum-) feeder[1].

Organism			Environment	
Molar char- acters	Character states (chewing adaptations)	bio- logi- cal role	physical pre- properties ferred preferred foods foods	
Cusps	*form* – very low relief, blunt, squat and rounded *function* – reduced penetrative function reflects great deformability of food	assist chemical digestion by producing greatest possible increase in surface area of ingested food	compact, 3-dimensional, watery, highly deformable (most mature pulp)	fruits (and gums) (accounting for nearly 100% of diet by weight per year)
Crests	*form* – greatly reduced in length, sharpness and salience; horizontal orientation and for occluding crests, reduced overall reciprocal curvature and differential orientation are characteristic *function* – point-cutting is greatly reduced because food is highly deformable, and can be more readily deformed between two broad surfaces			
Basins	*form* – rounded, virtually unconfined mesio-distally and very shallow, with surfaces that are weakly convexo-concave, and basically oriented normal to chewing force *function* – provide most efficient means of rapidly deforming a soft, watery mass of food by simply 'mashing' (crushing and grinding) food between two low relief surfaces. Abundant lateral food escapement assures maximum food deformation with minimum of food resistance			

natural selection – favors most efficient interaction between organism and environment – given heritage

[1] This model assumes a diet which is very restrictive. Further explanations are given in 'Hypotheses'.

Table III. Model of the molar adaptations of a (noncrosslophed) leaf-feeder[1].

Organism				Environment	
molar characters	character states (chewing adaptations)	biological role	⟷	physical properties preferred foods	preferred foods

Cusps	*form* – moderate relief, squat and elongated. Primitive, noncrosslophed configuration *function* – with crests they carry, distribute greatest possible length of sheet-like, compliant food over as great a length of cutting edge as possible. Leaf-pleating is poorly to moderate developed and transversely reciprocal			
Crests	*form* – their orientations strongly emphasize both horizontal and mesio-distal components, while occluding crests are reciprocally curved and/or differentially oriented primarily in a plane parallel with that of the crown cervix *function* – permit high pressure 'horizontal' point-cutting to be carried out over greatest possible length of tough, sheet-like, compliant food by distributing the greatest possible length of this food over the greatest possible length of cutting edges. Also contribute toward reduction of food resistance during point-cutting by permitting more vertical food escapement over a relatively greater horizontal distance, normal to chewing force	assist chemical digestion by producing greatest possible increase in surface area of ingested food	*natural selection* – favors most efficient interaction between organism and environment – given heritage	sheet-like, 2-dimensional, tough, fibrous, very compliant	leaves (accounting for nearly 100% of diet by weight per year)
Basins	*form* – mesio-distally broad and unconfined, but bucco-lingually steeply sloped, with planar surfaces oriented normal to the chewing force *function* – reduce food resistance during point cutting by facilitating vertical food escapement over relatively great horizontal distance, normal to chewing force. Permit maximum possible chewing force to be applied to greatest area of tough, sheet-like, compliant food thus facilitating crushing and grinding. During crushing and grinding, food resistance is reduced by permitting food escapement laterally, normal to chewing force			

[1] This model assumes a diet which is very restrictive. Further explanations are given in 'Hypotheses'.

Table IV. Model of the molar adaptations of a (crosslophed) leaf-leeder[1].

Organism					Environment	
molar characters	character states (chewing adaptations)	biological role	↔		physical properties preferred foods	preferred foods
Cusps	*form* – moderate to high relief, moderately tapered and pointed. Crosslophed configuration *function* – high relief cusps support a greater length of cutting edges. Crosslophing, however, permits greatest benefit to be derived from this fact by throwing tough, sheet-like, compliant food into deep, transversely identical folds or pleats which appose the greatest possible length of food to the greatest possible length of cutting edges	assist chemical digestion by producing greatest possible increase in surface area of ingested food	natural selection – favors most efficient interaction between organism and environment – given heritage	sheet-like, 2-dimensional, tough, fibrous, very compliant	leaves (accounting for nearly 100% of diet by weight per year)	
Crests	*form* – their orientations tend to emphasize components variably oblique to both the cervical plane and the mesial-distal axis of the molar, while occluding crests are reciprocally curved and/or differentially oriented primarily in a plane oblique to that of the crown cervix *function* – with help of pleating produced by crosslophing, permit high pressure point-cutting, intermediate between that of 'horizontal' and 'vertical', to be carried out over greatest possible length of tough, sheet-like, compliant food. Extensive food escapement in several directions minimizes food resistance during point-cutting					
Basins	*form* – replaced by mesio-distally convex and only moderately wide crosslophs with narrow subplanar slopes, near edges of basins, oriented oblique to chewing force *function* – primarily assist in pleating food to facilitate point-cutting. Have inefficient crushing and grinding capabilities					

[1] This model assumes a diet which is very restrictive. Further explanations are given in 'Hypotheses'.

Table V. Model of the molar adaptations of a stem-feeder[1].

Organism		biological role	↔	Environment	
molar characters	character states (chewing adaptations)			physical properties preferred foods	preferred foods
Cusps	*form* – moderate to high relief, squat and conical *function* – in combination with nature of occluding molar elements, permit moderately high pressured puncturing and extensive bending of tough, tubular food by virtue of tight, occlusal interdigitation of cusps, crests and basins of exaggerated differential relief	assist chemical digestion by producing greatest possible increase in surface area of ingested food	natural selection – favors most efficient interaction between organism and environment – given heritage	3-dimensional material with tubular, tough, fibrous, moderately compliant exterior and cylindrical, highly deformable pith	stems (accounting for nearly 100% of diet by weight per year)
Crests	*form* – their orientations moderately emphasize components both parallel with and normal to the cervical plane of the molar as well as a component oblique to the mesio-distal axis. Occluding crests are reciprocally curved and/or differentially oriented primarily in a plane oblique to that of the crown cervix. Total crest length is reduced by localized gaps in the basin walls *function* – permit thick, deformable pith of food to be 'punched out'. Allow point-cutting of moderate pressure and intermediate between that of 'horizontal' and 'vertical', to be carried out through a greater thickness of the tough, tubular exterior of food. Permit moderate reduction of food resistance during point-cutting by allowing some horizontal food escapement over a greater vertical distance, normal to chewing force. Some cutting function is compromised by relatively reduced crest length				
Basins	*form* – deep, rounded and mostly confined with convexo-concave to reciprocally concave surfaces oriented mostly oblique to chewing force. Possess large but localized gaps *function* – assist puncturing of tough, tubular food by restricting most lateral food escapement, while facilitating bending and 'punching' of food by adding to differential relief of molars. Localized gaps provide channel for lateral food escapement, thus controlling food resistance during puncturing. Efficient crushing and grinding are precluded				

[1] This model assumes a diet which is very restrictive. Further explanations are given in 'Hypotheses'.

III. Methods

A. Samples

The samples used in this investigation of strepsirhine molar adaptations consisted of dried skulls with the entire secondary dentition fully erupted. All extant material was from the collection of the Department of Mammalogy at the American Museum of Natural History. The subfossil material was from the collection of the Department of Vertebrate Paleontology at the same institution. Where possible, the left upper and lower second molars were measured. If these teeth were damaged, the contralateral teeth were used.

For the interspecific comparison of molar adaptations relating to diet, species samples were chosen which displayed relatively little and roughly comparable molar wear (or dentine exposure). Where possible, this entailed eliminating individuals whose values for Index XXV were greater than 0.25. [It was found that the index for wear in the $M_{\overline{2}}$ (Index XXV) was a more sensitive and more reliable indicator of overall molar wear than the index for wear in the $M^{\underline{2}}$ (Index V).]

Of the many specimens examined, the following species samples were used. Numbers in parentheses following species indicate sample size: *Lemur fulvus albifrons* (4), *Lemur mongoz mongoz* (3), *Lemur catta* (2), *Varecia variegatus* (3), *Lepilemur mustelinus leucopus* (4), *Hapalemur griseus olivaceous* (4), *Megaladapis edwardsi* (1), *Cheirogaleus major* (1), *Cheirogaleus medius* (1), *Microcebus murinus* (3), *Phaner furcifer* (2), *Galago crassicaudatus monteiri* (4), *Galago senegalensis moholi* (4), *Galago alleni* (2), *Galagoides demidovii anomurus* (4), *Euoticus elegantulus* (3), *Arctocebus calabarensis* (1), *Perodicticus potto ibeanus* (3), *Loris tardigradus* (4), *Nycticebus coucang javanicus* (3), *Indri indri* (3), *Propithecus verreauxi* (3), *Avahi laniger* (1), *Palaeopropithecus ingens* (1), *Archaeolemur majori* (1), *Hadropithecus stenograthus* (1), *Daubentonia madagascariensis* (1).

To investigate ontogenetic changes in molar wear and molar orientation in the strepsirhines, species samples were chosen which demonstrated dentine exposure ranging from little or none (with very low values for Index XXV) to very great (with very high values for Index XXV). The following species samples were utilized. Numbers in parentheses following species indicate sample size: *Lemur fulvus albifrons* (12), *Lepilemur mustelinus leucopus* (11), *Hapalemur griseus olivaceous* (7), *Galago crassicaudatus monteiri* (12), *Galago senegalensis moholi* (12), *Galagoides demidovii anomurus* (12). It was assumed that dental wear reflected chronological age.

B. Measurement Techniques

Guided by my hypotheses, the following measurements were taken on the left upper and lower second molars. Dental nomenclature is that used by van Valen (1966), Szalay [1969] and Kay [in press].

1. Linear Measurements

For most species, magnified drawings were made of appropriate views of the second molars with the aid of a Wild M-5 stereoscope and camera lucida, with reference lines superimposed on the drawings, where necessary. A set of Helios calipers, accurate to within $1/20$ mm, were then used to measure linear intervals on these drawings. These raw measurements were then corrected (standardized) for magnification. Views of the second molars were considered in the occlusal plane when the entire outline of the crown cervix was in a plane normal to the visual axis. Linear measurements were in millimeters. Helios calipers were used to take measurements directly from the large subfossils studied. Those measurements which are in parentheses are for indrioid strepsirhines.

a) Measurements of Basic Molar Proportions (in Occlusal Plane)

Length M^2 (fig. 7A): Distance between points of intersection of crown outline and line drawn through apices of paracone and metacone. Molar length = A–B.

Width M^2 (fig. 7A): Distance between points of intersection of crown outline and line drawn perpendicular to line A–B and through apex of protocone. Molar width = C–D.

Width of buccal portion of M^2 (fig. 7B): Maximum distance between points of intersection of buccal and lingual limits of paracone and line drawn through apex of paracone and perpendicular to reference line A–B. Width of buccal portion of M^2 = E–F.

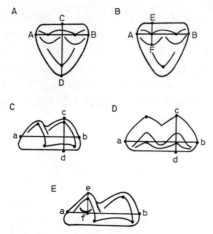

Fig. 7. Measurements of basic molar dimensions. For explanations, see text. *A* length and width of M^2. *B* Width of buccal portion of M^2. *C* Length and width of M$_{\bar{2}}$. *D* Length and width of M$_{\bar{2}}$ (for indrioids). *E* Width of buccal portion of M$_{\bar{2}}$. All drawings are occlusal views with buccal toward the top and mesial to the left.

Length M$_{\bar{2}}$ (fig. 7C): Distance between points of intersection of crown outline and line drawn through gingival-most points on protocristid and postcristid. Molar length = a–b.

[Length M$_{\bar{2}}$ (fig. 7D): Distance between points of intersection of crown outline and line drawn through apices of metaconid and entoconid. Molar length = a–b.]

Width M$_{\bar{2}}$ (fig. 7C): Distance between points of intersection of crown outline and line drawn perpendicular to line a–b and through apex of hypoconid. Molar width = c–d.

Width of buccal portion of M$_{\bar{2}}$ (fig. 7C): Maximum distance between points of intersection of buccal and lingual limits of protoconid and line drawn through apex of protoconid and perpendicular to reference line a–b. Width of buccal portion of M$_{\bar{2}}$ = e–f.

b) Measurements of Crest Length

All crests were viewed in a plane in which the greatest length of the crest was projected. The magnified crest was then drawn, using a camera lucida, and divided into nearly linear segments. These segments were measured with a set of Helios calipers and the segment measurements summed to obtain the total length of the crest. A crest is defined as a leading enamel cutting edge.

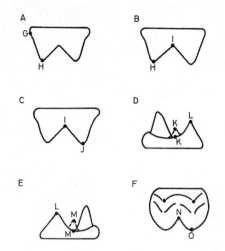

Fig. 8. Crest lengthⅰn the M². For explanations, see text. *A* Length of preparacrista (for indrioids). *B* Length of postparacrista. *C* Length of premetacrista. *D* Length of preprotocrista. *E* Length of postprotocrista. *F* Length of prehypocrista. *A–C* are buccal views with mesial to the left and dorsal toward the top; *D* is a mesial view with buccal to the left and ventral toward the top; *E* is a distal view with lingual to the left and ventral toward the top; *F* is occlusal view with mesial to the left and buccal toward the top.

Length preparacrista (fig. 8A): Length of crest between most gingival point on mesial end of ectoloph crest and apex of paracone. Length of preparacrista = G–H.

Length postparacrista (fig. 8B): Length of crest between most gingival point of centrocrista and apex of paracone. Length of postparacrista = H–I.

Length premetacrista (fig. 8C): Length of crest between most gingival point on centrocrista and apex of metacone. Length of premetacrista = I–J.

Length preprotocrista (fig. 8D): Length of crest between apex of proto-cone and either (a) most gingival point on crest between apex of protocone and buccal end of mesial cingulum; or (b) apex of paraconule. Length of preprotocrista = K–L.

Length postprotocrista (fig. 8E): Length of crest between apex of protocone and either (a) most gingival point on crest between apex of protocone and buccal end of distal cingulum; or (b) apex of metaconule. Length of postprotocrista = L–M.

[Length of prehypocrista (fig. 8F): Length of crest between apex of hypocone and point of junction with postprotocrista. Length of prehypo-crista = N–O.]

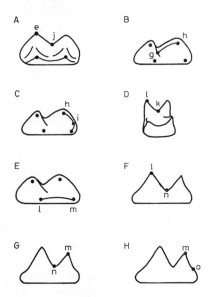

Fig. 9. Crest length in the M_3. For explanations, see text. *A* Length of protocristid (for indrioids). *B* Length of cristid obliqua. *C* Length of postcristid. *D* Length of (lingual portion of) protocristid. *E* Length of entocristid. *F* Length of postmetacristid (for indrioids). *G* Length of pre-entocristid (for indrioids). *H* Length of postentocristid (for indrioids). *A–C* and *E* are occlusal views with mesial to the left and buccal toward the top; *D* is a distal view with lingual to the left and dorsal toward the top; *F–H* are lingual views with mesial to the left and dorsal toward the top.

[Length of protocristid (fig. 9A): Length of crest between apex of protoconid and junction with cristid obliqua. Length protocristid = e–j].

Length cristid obliqua (fig. 9B): Length of crest bewteen apex of hypoconid and distal slope of protocristid. Length of cristid obliqua = g–h.

Length postcristid (fig. 9C): Length of crest between apex of hypoconid and most gingival point on crest between apex of hypoconid and apex of entoconid. Length of postcristid = h–i.

Length of (lingual portion of) protocristid (fig. 9D): Length of crest between apex of metaconid and most gingival point on crest between protoconid and metaconid. Length of lingual portion of protocristid = k–l.

Length entocristid (fig. 9E): Length of crest between apex of metaconid and apex of entoconid. Length of entocristid = l–m.

[Length postmetacristid (fig. 9F): Length of crest between apex of metaconid and most gingival point on talonid notch. Length postmetacristid = l–n.]

[Length pre-entocristid (fig. 9G): Length of crest between apex of entoconid and most gingival point on talonid notch. Length of pre-entocristid = m–n.]

Length post-entocristid (fig. 9H): Length of crest between apex of entoconid and distal rim of molar crown. Length of post-entocristid = m–o.]

c) Measurements of Crown Height and Cusp Proportions

The molar crown was viewed buccally, in a plane parallel to the cervical plane of the tooth. Using a camera lucida, magnified drawings of the molar crown were made, with appropriate reference lines superimposed. Helios calipers were then used to measure the linear intervals.

Crown height of M^2 (fig. 10A): Distance between apex of paracone and point on dentine-enamel junction lying on line drawn perpendicular to another line connecting mesial- and distal-most points on ectoloph. Crown height of M^2 = R–S.

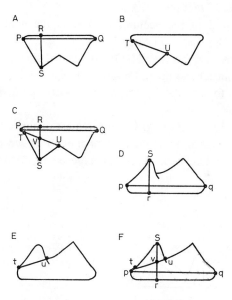

Fig. 10. Height of crown and cusp proportions in the M^2 and $M_{\overline{3}}$. For explanations, see text. *A* Height of crown of M^2. *B* Width of paracone. *C* Height of paracone. *D* Height of crown of $M_{\overline{3}}$. *E* Width of protoconid. *F* Height of protoconid. *A–C* are buccal views with mesial to the left and dorsal toward the top. *D–F* are buccal views with mesial to the left and dorsal toward the top.

Width of paracone (fig. 10B): Distance between most gingival point on centrocrista and most gingival point on preparacrista. Width of paracone = T–U.

Height of paracone (fig. 10C): Distance along reference line R–S between its intersections with apex of paracone and reference line T–U. Height of paracone = S–V.

Crown height of $M_{\overline{2}}$ (fig. 10D): Distance between apex of protoconid and point on dentine-enamel junction lying on line drawn perpendicular to another line connecting mesial- and distal-most points on ectolophid. Crown height of $M_{\overline{2}}$ = r–s.

Width of protoconid (fig. 10E): Distance between most gingival point at intersection of distal slope of protoconid and cristid obliqua and most gingival point on mesial slope of protoconid. Protoconid width = t–u.

Height of protoconid (fig. 10F): Distance along reference line r–s between its intersections with apex of protoconid and reference line t–u. Height of protoconid = s–v.

d) Measurements of Exposed Dentine

The molar crown was viewed in the occlusal plane and the dentine considered was at the apices of the cusps indicated below.

Width of exposed paracone dentine (fig. 11A): Greatest distance between points of intersection of buccal and lingual borders of exposed dentine and line drawn perpendicular to line A–B. Width of exposed paracone dentine = W–X.

Width of exposed protoconid dentine (fig. 11B): Greatest distance between points of intersection of buccal and lingual borders of exposed dentine and line drawn perpendicular to line a–b. Width of exposed protoconid dentine = w–x.

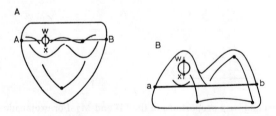

Fig. 11. Width of exposed dentine in the M^2 and $M_{\overline{2}}$. For explanations see text. *A* Width of exposed paracone dentine. *B* Width of exposed protoconid dentine. Both drawings are occlusal views with mesial to the left and buccal toward the top.

e) Measurement of (Reciprocal) Crest Curvature

Crest curvature was defined and measured in two planes as follows-

'Horizontal' – for the M^2, the curvature of the preprotocrista in a plane parallel to the buccal slope of protocone; for the $M_{\bar{3}}$, the curvature of protocristid (or for the indrioids, the entocristid) in a plane parallel to the cervical plane of the molar crown.

'Vertical' – for the M^2, the curvature of the preprotocrista in a plane perpendicular to the buccal slope of the protocone; for the $M_{\bar{3}}$, the curvature of the protocristid (or for the indrioids, the entocristid) in a plane perpendicular to the cervical plane of the molar crown.

A crest is defined as a leading enamel cutting edge.

Camera lucida drawings with appropriate superimposed reference lines were made, and Helios calipers used to take measurements from these drawings.

'Horizontal' curvature of the preprotocrista (fig. 12A): Greatest perpen-

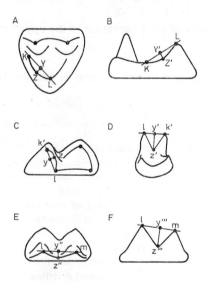

Fig. 12. Measurements of reciprocal crest curvature in the M^2 and $M_{\bar{3}}$. For explanations see text. A 'horizontal' curvature of preprotocrista. B 'vertical' curvature of preprotocrista. C 'horizontal' curvature of protocristid. D 'vertical' curvature of protocristid. E 'horizontal' curvature of entocristid (for indrioids). F 'vertical' curvature of entocristid (for indrioids). A, C and E are occlusal views with mesial to left and buccal toward the top; B is a mesial view with buccal to the left and ventral toward the top; D is a distal view with lingual to the left and dorsal toward the top; F is a lingual view with mesial to the left and dorsal toward the top.

dicular deviation of preprotocrista from chord connecting terminal points of this crest. Horizontal curvature of preprotocrista $= Y-Z$.

'Vertical' curvature of preprotocrista (fig. 12B): Greatest perpendicular deviation of preprotocrista from chord connecting terminal points of this crest. Vertical curvature of preprotocrista $= Y'-Z'$.

'Horizontal' curvature of protocristid (fig. 12C): Greatest perpendicular deviation of protocristid from chord connecting apices of protoconid and metaconid. Horizontal curvature of protocristid $= y-z$.

'Vertical' curvature of the protocristid (fig. 12D): Greatest perpendicular deviation of protocristid from chord connecting apices of protoconid and metaconid. Vertical curvature or protocristid $= y'-z'$.

['Horizontal' curvature of the entocristid (fig. 12E): Greatest perpendicular deviation of entocristid from chord connecting apices of metaconid and entoconid. Horizontal curvature of entocristid $= y''-z''$.]

'Vertical' curvature of the entocristid (fig. 12F): Greatest perpendicular deviation of entocristid from chord connecting apices of metaconid and entoconid. Vertical curvature of entocristid $= y'''-z'''$.

Signs used to indicate the direction of crest curvature in the upper and lower second molars are given in table VI.

Table VI. Signs used to indicate the direction of crest curvature in the upper and lower second molars[1].

M^2	$M_{\bar{3}}$
Preprotocrista	*Protocristid*
'Horizontal' curvature	'Horizontal' curvature
$+$ = mesio-lingual deviation	$+$ = disto-buccal deviation
$-$ = disto-buccal deviation	$-$ = mesio-lingual deviation
'Vertical' curvature	'Vertical' curvature
$+$ = gingival deviation	$+$ = gingival deviation
$-$ = apical deviation	$-$ = apical deviation
	(*Entocristid*
	'Horizontal' curvature
	$+$ = buccal deviation
	$-$ = lingual deviation
	'Vertical' curvature
	$+$ = gingival deviation
	$-$ = apical deviation)

[1] The term deviation refers to the deviation of a crest in a given plane from a chord connecting the terminal points of this crest. Curvature of the entocristid was considered in place of that of the protocristid for the Indriidae. Further explanations are given in the text.

2. Angular Measurements

a) Angles Formed between the Pre- and Postprotocrista
and between the Cristid Obliqua and Postcristid

The upper and lower second molars were viewed under a camera lucida in planes perpendicular to the buccal slope of protocone and the lingual slope of hypoconid, respectively. Chords were drawn through the appropriate crests, and the angles between the intersecting chords were measured (in degrees) using a protractor.

Angle between pre- and postprotocrista (fig. 13A): Angle measured at apex of protocone between chords connecting protocone apex with terminal points of preprotocrista and postprotocrista. Angle between pre- and postprotocrista = α.

Angle between cristid obliqua and postcristid (fig. 13B): Angle measured at apex of hypoconid between chords connecting hypoconid apex with terminal points of cristid obliqua and postcristid. Angle between cristid obliqua and postcristid = β.

b) Angles Formed by the Trigon Basin and Talonid Basin Contours
along the Mesio-Distal and Bucco-Lingual Axes

The $M^{\underline{2}}$ and $M_{\overline{3}}$ were viewed lingually under the camera lucida. A cone of modelling clay with a reference line facing the visual axis was held at its blunt end by a pair of forceps, with the long axis of the clay cone perpendicular to both the visual axis and the cervical plane of the molar. With the clay cone so oriented, the tapered end of the clay cone was pressed into the molar basin. The clay cone was then withdrawn from the molar basin and viewed. The mesio-distal profile of the clay mold of the molar basin was perpendicular to the visual axis, and the reference line faced the visual axis. Two chords were drawn with the aid of the camera lucida, connecting the gingival-most point on the basin mold with the mesial and distal rims of the basin mold, respectively. A protractor was then used to measure the angle (in degrees) of the mesio-distal contour of the molar basin by measuring the angle formed by the two chords as they intersected the gingival-most point of the basin mold. With the long axis of the clay cone still perpendicular to the visual axis, the clay cone was rotated 90 degrees (with the aid of the reference line) and the angle of the bucco-lingual contour of the molar basin was obtained in similar fashion. Angles ranging between 0 and 179 degrees indicate a concave basin, an angle of 180 degrees indicates a flat basin, while angles above 180 degrees indicate a convex 'basin'.

Angle formed by the mesio-distal contour of the trigon basin (fig. 13C):

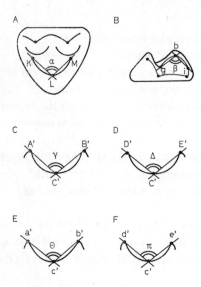

Fig. 13. Angular measurements of the configurations of the trigon and talonid basins. For explanations, see text. *A* Angle between the pre- and postprotocrista. *B* Angle between the cristid obliqua and postcristid. *C* Angle of the mesio-distal contour of the trigon basin. *D* Angle of the bucco-lingual contour of the trigon basin. *E* Angle of the mesio-distal contour of the talonid basin. *F* Angle of the bucco-lingual contour of the talonid basin. *A* and *B* are occlusal views with mesial to the left and buccal toward the top; *C* is a mesio-distal section of the trigon basin with ventral toward the top; *D* is a bucco-lingual section of the trigon basin with ventral toward the top; *E* is a mesio-distal section of the talonid basin with dorsal toward the top; *F* is a bucco-lingual section of the talonid basin with dorsal toward the top.

Angle formed between chords connecting the gingival-most point on the mesio-distal profile of basin mold with, respectively, the mesial and distal rims of the mesio-distal profile of the basin mold. Angle of mesio-distal contour of trigon basin = γ.

Angle formed by the bucco-lingual contour of the trigon basin (fig. 13D): Angle formed between chords connecting the gingival-most point on the bucco-lingual profile of basin mold with, respectively, the buccal and lingual rims of the bucco-lingual profile of the basin mold. Angle of bucco-lingual contour of trigon basin = Δ.

For the talonid basin, the angles of the mesio-distal and bucco-lingual contours of the talonid basin are defined similarly as those of the trigon basin. The angle of the mesio-distal contour of the talonid basin (fig. 13E) = θ. The angle of the bucco-lingual contour of the talonid basin (fig. 13F) = π.

(For indrioids, the trigon basin is considered the area bounded by the pre- and postprotocrista; the talonid basin is considered the area bounded by the cristid obliqua and postcristid.)

c) Orientation of Buccal Phase Facetting and Molar Torsion
(for M^2 only)

A thin thread was made into a noose and tightened at equivalent points across the palate, as near as possible to the M^2. This thread was considered parallel with the palatal plane. The M^2 was viewed distally (as parallel as possible to the cervical plane of the tooth) through the camera lucida, with the thread running across the upper portion of the visual field. Several lines were then drawn as follows (fig. 14): line 1, drawn over the image of the thread; line 2, drawn parallel to the buccal phase facets (or facet striations); line 3, drawn between the most buccal and the most lingual points visible on the crown cervix; line 4, drawn parallel to line 1, but beneath the image of the molar, and intersecting lines 2 and 3.

A protractor was then used to measure the following angles (in degrees and in the coronal plane):

Angle Φ = angle between lines 2 and 4 = angle buccal phase facetting makes with palatal plane.

Angle μ = angle between lines 3 and 4 = angle of molar torsion.

Angle ε = angle between lines 2 and 3 = angle buccal phase facetting makes with cervical plane of molar.

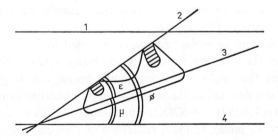

Fig. 14. Orientation of buccal phase facetting and molar torsion (for the M^2 only). For explanations, see text. Figure shows measurements taken on left upper second molar in distal view, with lingual to the left and ventral toward the top. A positive value for μ indicates a lingual tilt to the molar, while a negative value for μ indicates a buccal tilt to the molar.

C. Indices of Molar Form and Function

In table VII, the measurements taken on the M^2 and $M_{\overline{3}}$ are organized into indices. The morphological and functional significance of these indices is indicated below. When comparable, indices for M^2 and $M_{\overline{3}}$ are treated together. These indices were used in interspecific statistical comparisons to investigate the nature of species-specific molar adaptations, and in both intra- and interspecific analyses to investigate patterns of ontogenetic wear and molar reorientation within the strepsirhines.

Indices I and XXI and II and XXII reflect molar area and molar size, while Indices III and XXIII compare molar length and molar width and reflect basic molar proportions. The morphological predominance of the buccal portion of the molar is indicated by Indices IV and XXIV, while Indices V and XXV reflect cusp truncation or molar wear, with Index XXV being the more sensitive and reliable indicator of overall wear.

Indices VI and XXVI are indicators of mesio-distal cusp acuity as well as of the degree to which the orientations of the crest edges along the ectoloph and ectolophid emphasize a component perpendicular to the cervical plane of the molar. These indices reflect the emphasis on point penetration and 'vertical' point-cutting. Indices VII and XXVII indicate cusp relief as well as the degree to which the orientations of the crest edges along the ectoloph and ectolophid emphasize a component perpendicular to the cervical plane of the molar. They reflect the emphasis on both point penetration and 'vertical' point-cutting. Bucco-lingual cusp acuity is indicated by Indices VIII and XXVIII which reflect the emphasis on point penetration, while Indices IX and XXIX indicate crown height and presumably reflect the amount of crown material which can sustain ontogenetic wear, and hence, presumably reflect the functional longevity of the molar.

The relative development of molar cresting is indicated by Indices X and XXX, which reflect the total emphasis on point-cutting, while Indices XI and XXXI compare the length of cresting in the buccal and lingual portions of the molar, and may reflect the emphasis on point-cutting in one of these two portions. Indices XII and XXXII compare the lengths of crests in the lingual portion of the molar, and may reflect not only an emphasis on prevallum or postvallum point-cutting but also unusual escapement patterns in the trigon and talonid basins.

Indices XIII and XXXIII indicate the component of the reciprocal edge curvature of the preprotocrista and protocristid (or entocristid) which lies in a plane roughly parallel with that of the crown cervix, and reflect the empha-

Table VII. Indices of molar form and function[1].

M^2	$M_{\bar{2}}$
I molar length × width	XXI molar length × width
II $\sqrt{\text{molar length} \times \text{width}}$	XXII $\sqrt{\text{molar length} \times \text{width}}$
III molar length/width	XXIII molar length/width
IV $\dfrac{\text{width of buccal portion of molar}}{\text{total molar width}}$	XXIV $\dfrac{\text{width of buccal portion of molar}}{\text{total molar width}}$
V $\dfrac{\text{width of exposed paracone dentine}}{\text{width of buccal portion of molar}}$	XXV $\dfrac{\text{width of exposed protoconid dentine}}{\text{width of buccal portion of molar}}$
VI $\dfrac{\text{height of paracone}}{\text{width of paracone}}$	XXVI $\dfrac{\text{height of protoconid}}{\text{width of protoconid}}$
VII $\dfrac{\text{height of paracone}}{\sqrt{\text{molar length} \times \text{width}}}$	XXVII $\dfrac{\text{height of protoconid}}{\sqrt{\text{molar length} \times \text{width}}}$
VIII $\dfrac{\text{height of paracone}}{\text{width of buccal portion of molar}}$	XXVIII $\dfrac{\text{height of protoconid}}{\text{width of buccal portion of molar}}$
IX $\dfrac{\text{height of crown}}{\sqrt{\text{molar length} \times \text{width}}}$	XXIX $\dfrac{\text{height of crown}}{\sqrt{\text{molar length} \times \text{width}}}$
X $\dfrac{\text{sum of lengths of postparacrista, premetacrista, preprotocrista and postprotocrista}}{\sqrt{\text{molar length} \times \text{width}}}$	XXX $\dfrac{\text{sum of lengths of cristid obliqua, postcristid, protocristid, entocristid}}{\sqrt{\text{molar length} \times \text{width}}}$
(or for indriids)	
$\dfrac{\text{sum of lengths of preparacrista, postparacrista, premetacrista, preprotocrista, postprotocrista and prehypocrista}}{\sqrt{\text{molar length} \times \text{width}}}$	$\dfrac{\text{sum of lengths of protocristid, cristid obliqua, postcristid, postmetacristid, pre-entocristid and postentocristid}}{\sqrt{\text{molar length} \times \text{width}}}$
XI $\dfrac{\text{sum of lengths of postparacrista and premetacrista}}{\text{sum of lengths of preprotocrista and postprotocrista}}$	XXXI $\dfrac{\text{sum of lengths of cristid obliqua and postcristid}}{\text{sum of lengths of protocristid and entocristid}}$
(or for indriids)	
$\dfrac{\text{sum of lengths of preparacrista, postparacrista, and premetacrista}}{\text{sum of lengths of preprotocrista, postprotocrista, and prehypocrista}}$	$\dfrac{\text{sum of lengths of protocristid, cristid obliqua, and postcristid}}{\text{sum of lengths of postmetacristid, pre-entocristid and postentocristid}}$

Table VII (continued)

M²	M₂
XII length of postprotocrista / length of preprotocrista	XXXII length of entocristid / length of protocristid (or for indriids) length of pre-entocristid / length of postmetacristid
XIII 'horizontal' perpendicular deviation of preprotocrista / chord length of preprotocrista	XXXIII 'horizontal' perpendicular deviation of protocristid / chord length of protocristid (or for indriids) 'horizontal' perpendicular deviation of entocristid / chord length of entocristid
XIV 'vertical' perpendicular deviation of preprotocrista / chord length of preprotocrista	XXXIV 'vertical' perpendicular deviation of protocristid / chord length of protocristid (or for indriids) 'vertical' perpendicular deviation of entocristid / chord length of entocristid
XV angle between preprotocrista and postprotocrista	XXXV angle between the cristid obliqua and postcristid
XVI angle of mesio-distal contour of trigon basin	XXXVI angle of mesio-distal contour of talonid basin
XVII angle of bucco-lingual contour of trigon basis	XXXVII angle of bucco-lingual contour of talonid basin

For M² only (all in coronal plane)
XVIII angle between the axis of orientation of buccal phase facetting and the palatal plane
 XIX angle between the cervical plane of molar crown and the palatal plane
  XX angle between the axis of orientation of buccal phase facetting and the cervical plane of the molar crown

[1] Indices I–XX pertain to the upper molar while Indices XXI–XXXVII pertain to the lower molar. Comparable indices for the upper and lower molars are horizontally paired. The significance of each index is explained in the text.

sis on 'horizontal' point-cutting. Indices XIV and XXXIV, on the other hand, indicate the component of the reciprocal edge curvature of the above-mentioned crests which lies in a plane perpendicular to that of the crown cervix. These indices reflect the emphasis on 'vertical' point-cutting.

The mesio-distal narrowing of the trigon and talonid basins as well as the orientation of the crests around these basins are indicated by Indices XV and XXXV. These indices primarily reflect whether occluding molar crests are (a) reciprocally curved and/or differentially oriented in a plane which is both normal to that of the crown cervix and parallel with the bucco-lingual axis of the molar (i.e., primarily designed for 'vertical' point-cutting) or (b) reciprocally curved and/or differentially oriented primarily in a plane parallel with that of the crown cervix and oriented primarily mesio-distally (i.e., primarily designed for 'horizontal' point-cutting). These indices may also reflect the tendency to emphasize either point penetration or crushing and grinding in the above-mentioned basins. Indices XVI and XXXVI indicate the mesio-distal concavity and confinement of the trigon and talonid basins, as well as the relief of the crests surrounding these basins. These indices reflect: whether point penetration (and attendant escapement patterns) or crushing and grinding (and attendant escapement patterns) are emphasized in the above-mentioned basins; whether food pleating is emphasized; and whether 'vertical' or 'horizontal' point-cutting (and attendant escapement patterns) is emphasized along the crests surrounding the trigon and talonid basins. The bucco-lingual concavity of the trigon and talonid basins and the inclination of the incusion surfaces in these basins are indicated by Indices XVII and XXXVII. These indices primarily reflect the orientation of the basin incusion surfaces with respect to the cervical plane of the molar (and the chewing force).

Index XVIII reflects the orientation of the trajectory of the lower second molar during buccal phase chewing, with respect to the palatal plane, while the extent of molar torsion, or buccal eruption, with respect to the palatal plane is reflected by Index XIX. Index XX reflects the orientation of the trajectory of the lower second molar during buccal phase chewing with respect to the features of the crown of the upper second molar.

D. Predictions Based on Hypothetical Models

My hypotheses of strepsirhine molar adaptations predict that species values for representative indices of molar form will vary with respect to the hypothetical dietary preferences specified in a manner indicated in table VIII. Predicted values for the species means of each index considered are in the form of a rank ordering. Ranks are designated from 1 to 5, with 1 indicating the highest value for an index and 5 indicating the lowest value for an index. A number in parentheses indicates the rank of a lower second molar value if different from that of the upper second molar.

One can expect that actual species, with less restricted diets, will demonstrate relative index values intermediate between those indicated in table VIII, and within the appropriate interval of dietary preferences.

Table VIII. Predictions of relative values of species means for each hypothetical dietary preference[1].

Indices considered	Hypothetical dietary preferences				
	insects	fruits	leaves		stems
			(a)	(b)	
VI and XXVI	1	5	4 (2)	2 (4)	3
VII and XXVII	1	5	4 (2)	2 (4)	3
VIII and XXVIII	1	5	4	2	3
IX and XXIX	1	4	3	1	2
X and XXX	1	4	2	1	3
XIII + XXXIII ——————— XIV + XXXIV	5	1	2	4	3
XIV and XXXIV	1	5	4	2	3
XV and XXXV	4 (5)	2	1	3 (4)	5 (3)
XVI and XXXVI	5	2	3	1	4

[1] The table shows the expected pattern of variation in species means for several indices of molar form and function as this relates to the hypothetical dietary preferences indicated. Each dietary preference is nearly exclusive. Those forms which prefer leaves are divided into two groups: (a) pertains to leaf-feeders with noncrosslophed molars, (b) pertains to leaf-feeders with crosslophed molars. Predicted values for indices for each hypothetical dietary group are rank-ordered, with '1' indicating the highest, and '5' indicating the lowest value for an index. A number in parentheses indicates the rank of a second lower molar value if different from that of the upper second molar. Further explanations are given in the text.

E. Statistical Methods

To investigate and test hypotheses relating to species-specific molar adaptations related to diet, histograms were employed (fig. 15–43) to compare species means for relevant indices of molar form. The mean for each index of molar form was calculated for each species with the help of an Olivetti Underwood Programma 101 desk calculator.

The nature of the data in this study rendered the use of more sophisticated parametric statistical techniques unfeasible. Bivariate correlation, multiple correlation and factor analysis were precluded, for example, because of the nominal or ordinal nature of the independent variables, while the often high degree of interrelatedness of the independent variables prevented the use of analyses of covariance. The great many subsamples also made the use of Student t tests impractical.

Histogram analyses were carried out on representative indices of molar form to determine whether the differences in the species means for these indices were best explained by molar size, or heritage, or dietary preference. To accomplish this, three histogram analyses of each index of molar form were employed: (1) in the first histogram analysis, the species means for a particular index were arranged according to the increasing size of both the upper and lower second molars (i.e., according to the increasing values of Indices II and XXII); (2) in the second historgam analysis, species means were arranged according to their taxonomic grouping (i.e., by family or subfamily): Daubentoniidae, Indriidae, Lemurinae, Cheirogaleinae, Galaginae, Lorisinae (SZALAY and KATZ's [1973] model of strepsirhine phylogeny was assumed here); (3) in the third analysis, species means were arranged according to the known dietary preference of each species. The sequence of dietary preferences used as the framework within which the species means were arranged is given in figure 44A.

A comparison of the sequences of species means, arranged according to the three aforementioned criteria, permitted the determination of whether (a) molar size or (b) phylogeny or (c) dietary preference best predicted the values of the various indices of molar form.

Comparable indices for upper and lower molars were treated simultaneously, but on separate histograms. This approach revealed any discrepancies between patterns demonstrated by the upper and lower molars.

To investigate ontogenetic changes in molar wear and molar orientation in several relevant species samples, bivariate correlation and linear regression analyses were carried out. These permitted the determination within each species sample of the relationships between (a) molar wear (Index XXV), (b) the orientation of buccal phase facetting (Index XX), and (c) the degree of molar torsion (Index XIX). Pearson's r and regression lines were calculated with the aid of an Olivetti Underwood Programma 101 desk calculator. These analyses for *Lepilemur* are shown in figures 45–47.

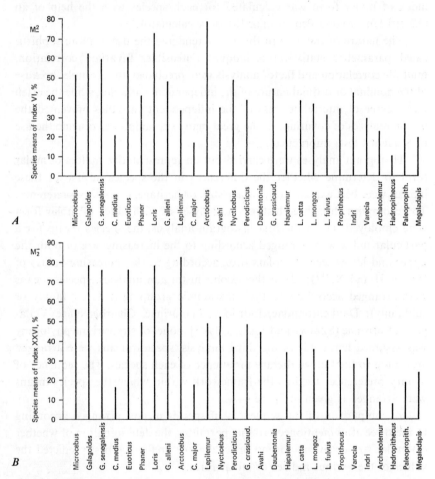

Fig. 15. Mesio-distal cusp acuity in the M2's as it varies with molar size. *A* compares the M² of species with respect to species means of Index VI (height of paracone/width of paracone). Species means are arranged according to upper molar size (Index II), with molar size increasing toward the right. *B* compares the M₂ of species with respect to species means of Index XXVI (height of protoconid/width of protoconid). Species means are arranged according to lower molar size (Index XXII), with molar size increasing toward the right. Higher values indicate greater cusp acuity.

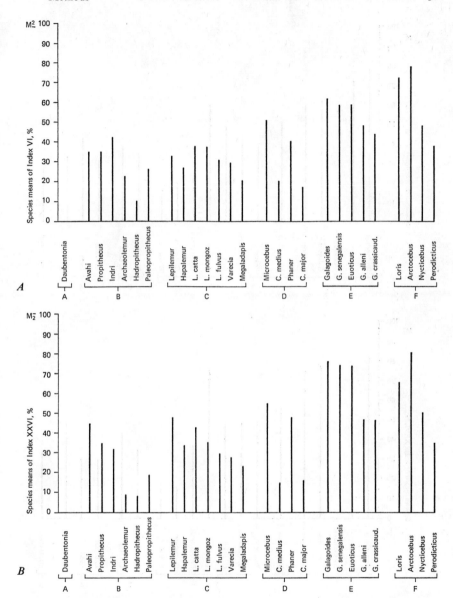

Fig. 16. Mesio-distal cusp acuity in the M2's as it varies taxonomically. Species means are grouped by family or subfamily: (A) Daubentoniidae; (B) Indriidae; (C) Lemurinae; (D) Cheirogaleinae; (E) Galaginae; (F) Lorisinae. *A* compares the M² of species with respect to species means of Index VI (height of paracone/width of paracone). *B* compares the M₂ of species with respect to species means of Index XXVI (height of protoconid/width of protoconid). Higher values indicate greater cusp acuity.

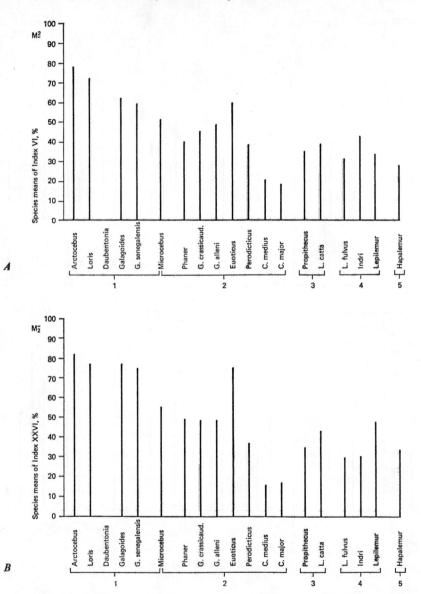

Fig. 17. Mesio-distal cusp acuity in the M 2's as it varies with dietary preference. Species means are arranged according to dietary preference in a manner explained in table IX. *A* compares the M2 of species with respect to species means of Index VI (height of paracone/width of paracone). *B* compares the M2 of species with respect to species means of Index XXVI (height of protoconid/width of protoconid). Higher values indicate greater cusp acuity.

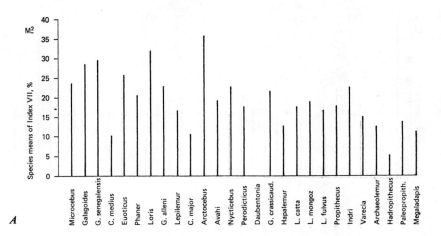

A

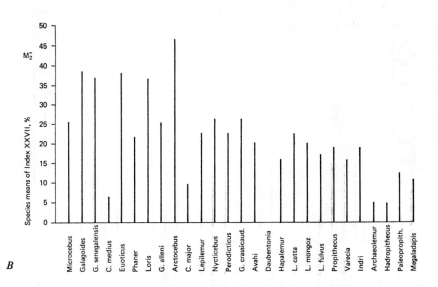

B

Fig. 18. Cusp relief in the second molars as it varies with molar size. *A* compares the M² of species with respect to species means of Index VII (height of paracone/$\sqrt{\text{molar length} \times \text{width}}$). Species means are arranged according to upper molar size (Index II), with molar size increasing toward the right. *B* compares the M$\bar{2}$ of species with respect to species means of Index XXVII (height of protoconid/$\sqrt{\text{molar length} \times \text{width}}$). Species means are arranged according to lower molar size (Index XXII), with molar. size increasing toward the right. Higher values indicate greater cusp relief.

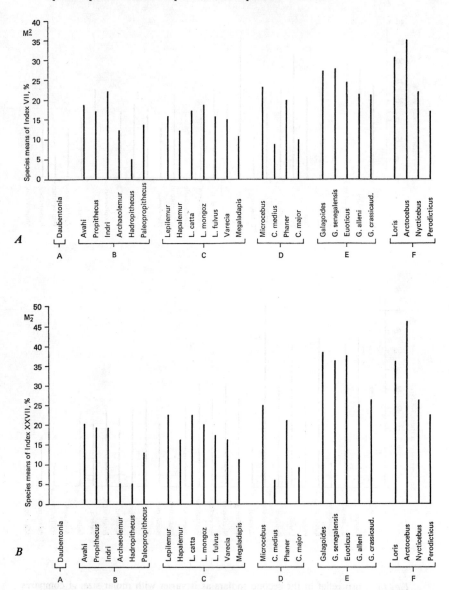

Fig. 19. Cusp relief in the second molars as it varies taxonomically. Species means are grouped by family or subfamily: (A) Daubentoniidae, (B) Indriidae; (C) Lemurinae; (D) Cheirogaleinae; (E) Galaginae; (F) Lorisinae. *A* compares the M² of species with respect to species means of Index VII (height of paracone/$\sqrt{\text{molar length} \times \text{width}}$). *B* compares the M$_{\overline{2}}$ of species with respect to species means of Index XXVII (height of protoconid/$\sqrt{\text{molar length} \times \text{width}}$). Higher values indicate greater cusp relief.

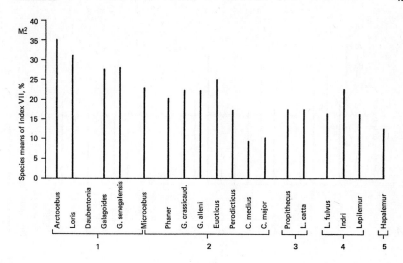

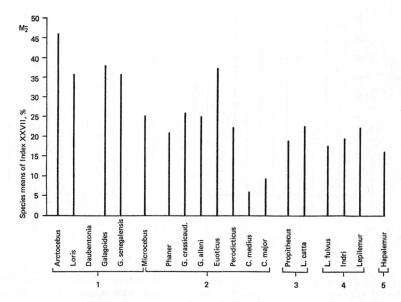

Fig. 20. Cusp relief in the second molars as it varies with dietary preference. Species means are arranged according to dietary preference in a manner explained in table IX. *A* compares the M^2 of species with respect to species means of Index VII (height of paracone/$\sqrt{\text{molar length} \times \text{width}}$). *B* compares the $M_{\bar{2}}$ of species with respect to species means of Index XXVII (height of protoconid/$\sqrt{\text{molar length} \times \text{width}}$). Higher values indicate greater cusp relief.

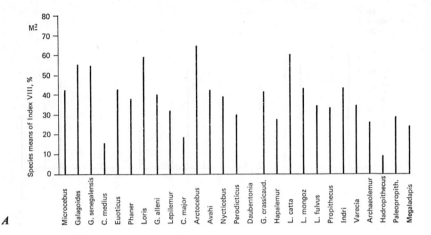

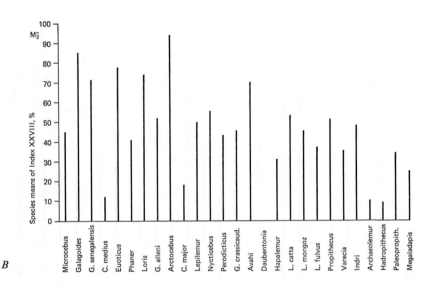

Fig. 21. Bucco-lingual cusp acuity in the M2's as it varies with molar size. *A* compares the M^2 of species with respect to species means of Index VIII (height of paracone/width of buccal portion of molar). Species means are arranged according to upper molar size (Index II), with molar size increasing toward the right. *B* compares the M$_{\overline{2}}$ of species with respect to species means of Index XXVIII (height of protoconid/width of buccal portion of molar). Species means are arranged according to lower molar size (Index XXII), with molar size increasing toward the right. Higher values indicate greater cusp acuity.

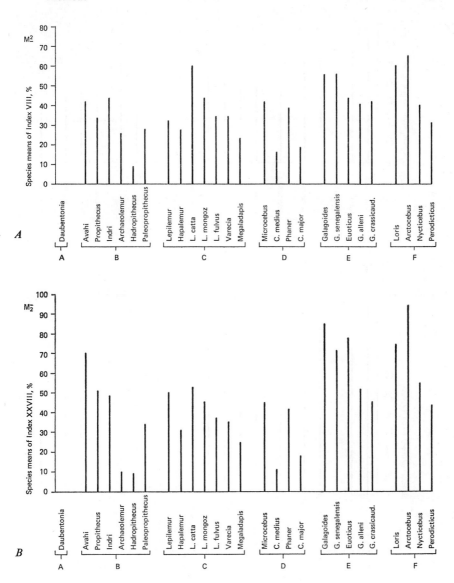

Fig. 22. Bucco-lingual cusp acuity in the M2's as it varies taxonomically. Species means are grouped by family or subfamily: (A) Daubentoniidae; (B) Indriidae; (C) Lemurinae; (D) Cheirogalainae; (E) Galaginae; (F) Lorisinae. *A* compares the M² of species with respect to species means of Index VIII (height of paracone/width of buccal portion of molar). *B* compares the M₂ of species with respect to species means of Index XXVIII (height of protoconid/width of buccal portion of molar). Higher values indicate greater cusp acuity.

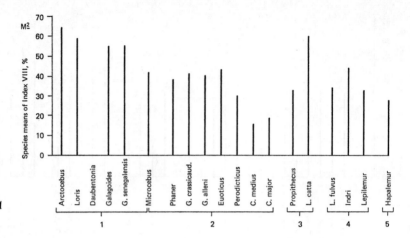

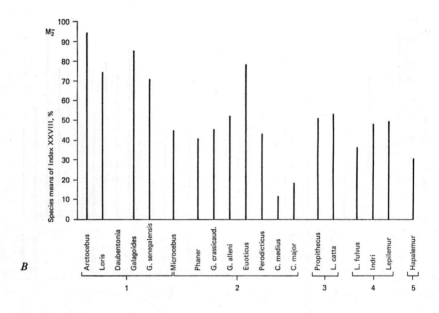

Fig. 23. Bucco-lingual cusp acuity in the M2's as it varies with dietary preference. Species means are arranged according to dietary preference in a manner explained in table IX. *A* compares the M$\overline{2}$ of species with respect to species means of Index VIII (height of paracone/width of buccal portion of molar). *B* compares the M$\overline{2}$ of species with respect to species means of Index XXVIII (height of protoconid/width of buccal portion of molar). Higher values indicate greater cusp acuity.

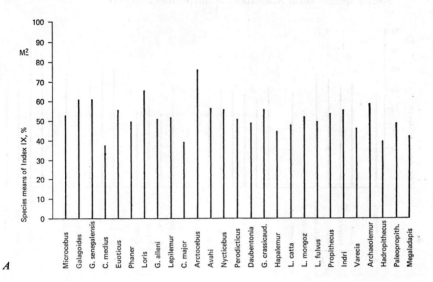

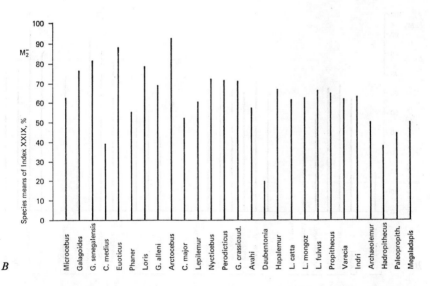

Fig. 24. Height of crown of the second molars as it varies with molar size. *A* compares the M² of species with respect to species means of Index IX (height of crown/ $\sqrt{\text{molar length} \times \text{width}}$). Species means are arranged according to upper molar size (Index II), with molar size increasing toward the right. *B* compares the M₂ of species with respect to species means of Index XXIX (height of crown/ $\sqrt{\text{molar length} \times \text{width}}$). Species means are arranged according to lower molar size (Index XXII), with molar size increasing toward the right. Higher values indicate greater crown height.

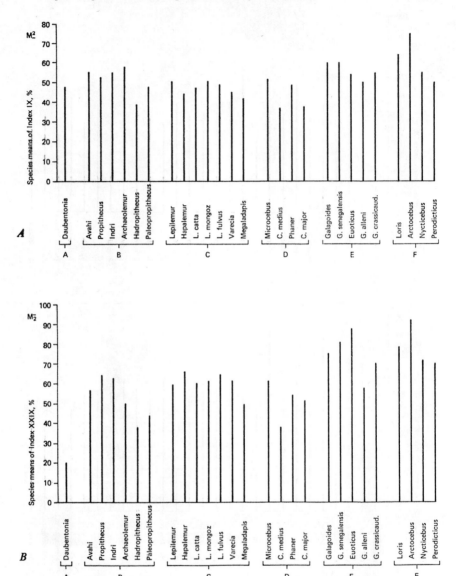

Fig. 25. Height of crown of the second molars as it varies taxonomically. Species means are grouped by family or subfamily: (A) Daubentoniidae; (B) Indriidae; (C) Lemurinae; (D) Cheirogaleinae; (E) Galaginae; (F) Lorisinae. *A* compares the M^2 of species with respect to species means of Index IX (height of crown/$\sqrt{\text{molar length} \times \text{width}}$). *B* compares the M$_2$ of species with respect to species means of Index XXIX (height of crown/$\sqrt{\text{molar length} \times \text{width}}$). Higher values indicate greater crown height.

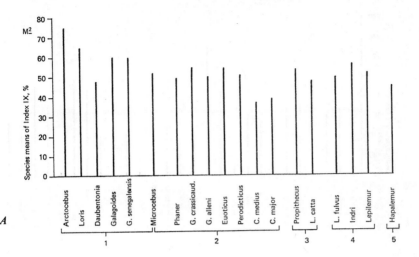

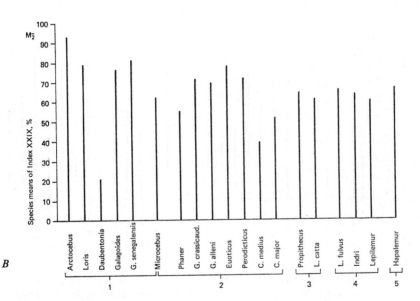

Fig. 26. Height of crown of the second molars as it varies with dietary preference. Species means are arranged according to dietary preference in a manner explained in table IX. *A* compares the M^2 of species with respect to species means of Index IX (height of crown/$\sqrt{\text{molar length} \times \text{width}}$). *B* compares the $M_{\bar{2}}$ of species with respect to species means of Index XXIX (height of crown/$\sqrt{\text{molar length} \times \text{width}}$). Higher values indicate greater crown height.

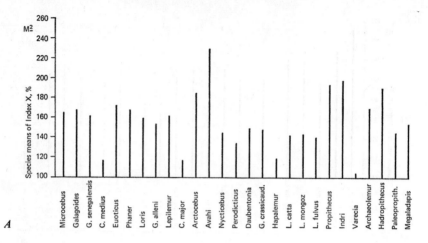

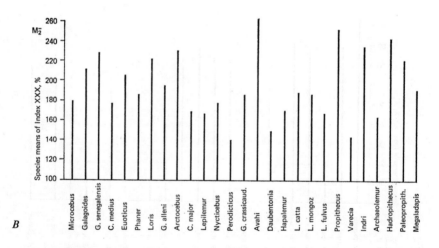

Fig. 27. Emphasis on crest development in the second molars as it varies with molar size. *A* compares the M^2 of species with respect to species means of Index X (sum of length of several molar crests/$\sqrt{\text{molar length} \times \text{width}}$). Species means are arranged according to upper molar size (Index II), with molar size increasing toward the right. *B* compares the M$_{\overline{2}}$ of species with respect to species means of Index XXX (sum of length of several molar crests/$\sqrt{\text{molar length} \times \text{width}}$). Species means are arranged according to lower molar size (Index XXII), with molar size increasing toward the right. Higher values indicate greater molar crest development.

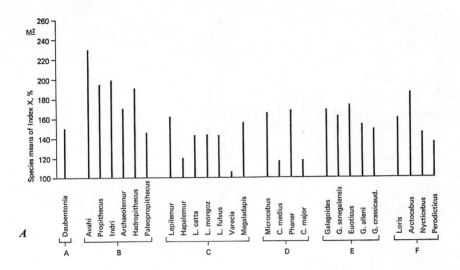

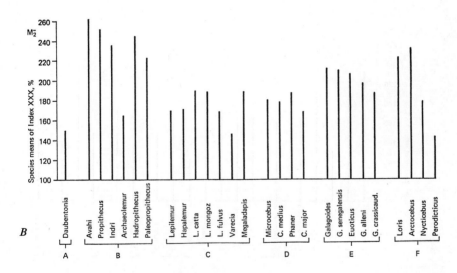

Fig. 28. Emphasis on crest development in the second molars as it varies taxonomically. Species means are grouped by family or subfamily: (A) Daubentoniidae; (B) Indriidae; (C) Lemurinae; (D) Cheirogaleinae; (E) Galaginae; (F) Lorisinae. *A* compares the M² of species with respect to species means of Index X (sum of lengths of several molar crests/$\sqrt{\text{molar length} \times \text{width}}$). *B* compares the M₂̄ of species with respect to species means of Index XXX (sum of lengths of several molar crests/$\sqrt{\text{molar length} \times \text{width}}$). Higher values indicate greater molar crest development.

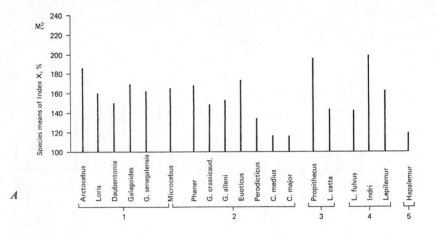

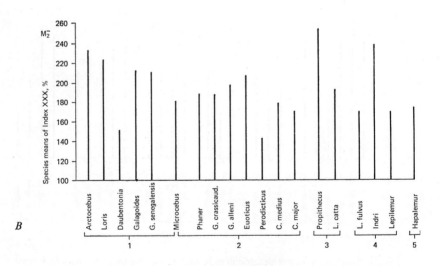

Fig. 29. Emphasis on crest development in the second molars as it varies with dietary preference. Species means are arranged according to dietary preference in a manner explained in table IX. *A* compares the M^2 of species with respect to species means of Index X (sum of lengths of several molar crests/$\sqrt{\text{molar length} \times \text{width}}$). *B* compares the $M_{\overline{2}}$ of species with respect to species means of Index XXX (sum of lengths of several molar crests/$\sqrt{\text{molar length} \times \text{width}}$). Higher values indicate greater molar crest development.

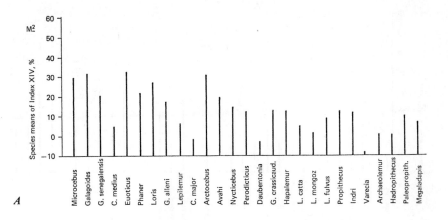

A

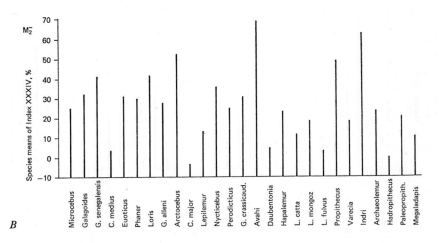

B

Fig. 30. 'Vertical' crest curvature in the second molars as it varies with molar size. *A* compares the M^2 of species with respect to species means of Index XIV ('vertical' perpendicular deviation of preprotocrista/chord length of preprotocrista). Species means are arranged according to upper molar size (Index II), with molar size increasing toward the right. *B* compares the M$_{\overline{2}}$ of species with respect to species means of Index XXXIV ['vertical' perpendicular deviation of protocristid (or entocristid)/chord length of protocristid (or entocristid)]. Species means are arranged according to lower molar size (Index XXII), with molar size increasing toward the right. Higher absolute values indicate greater 'vertical' curvature of molar crests. Positive and negative values refer to the directions of 'vertical' crest curvature. Table VI relates signs of values to the direction of crest curvature.

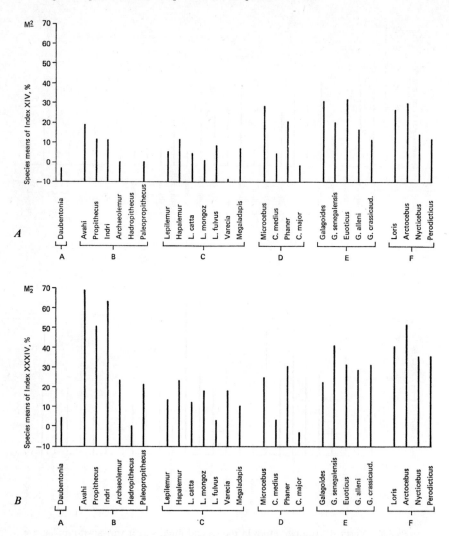

Fig. 31. 'Vertical' crest curvature in the second molars as it varies taxonomically. Species means are grouped by family or subfamily: (A) Daubentoniidae; (B) Indriidae; (C) Lemurinae; (D) Cheirogaleinae; (E) Galaginae; (F) Lorisinae. *A* compares the M² of species with respect to species means of Index XIV ('vertical' perpendicular deviation of preprotocrista/chord length of preprotocrista). *B* compares the M$_{\overline{2}}$ of species with respect to species means of Index XXXIV ['vertical' perpendicular deviation of proto-cristid (or entocristid)/chord length of protocristid (or entocristid)]. Higher absolute values indicate greater 'vertical 'curvature of molar crests. Positive and negative values refer to the directions of 'vertical' crest curvature. Table VI relates signs of values to the direction of crest curvature.

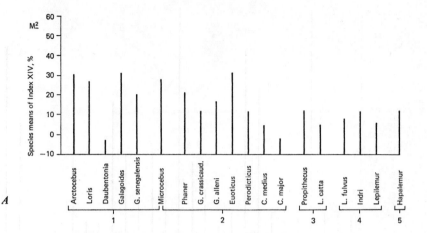

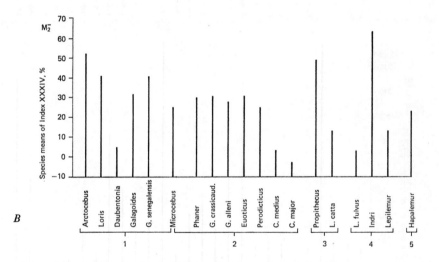

Fig. 32. 'Vertical' crest curvature in the second molars as it varies with dietary preference. Species means are arranged according to dietary preference in a manner explained
in table IX. *A* compares the M² of species with respect to species means of Index XIV
('vertical' perpendicular deviation lf preprotocrista/chord length of preprotocrista). *B*
compares the M₂ of species with respect to species means of Index XXXIV ['vertical'
perpendicular deviation of protocristid (or entocristid)/chord length of protocristid (or
entocristid)]. Higher absolute values indicate greater 'vertical' curvature of molar crests.
Positive and negative values refer to the directions of 'vertical' crest curvature. Table VI
relates signs of values to the direction of crest curvature.

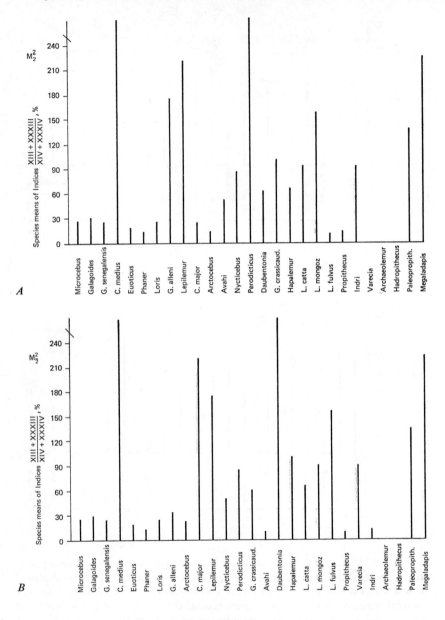

Fig. 33. For legend see page 61.

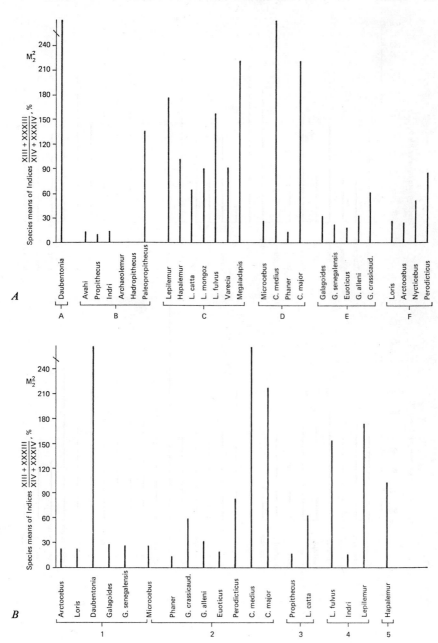

Fig. 34. For legend see page 61.

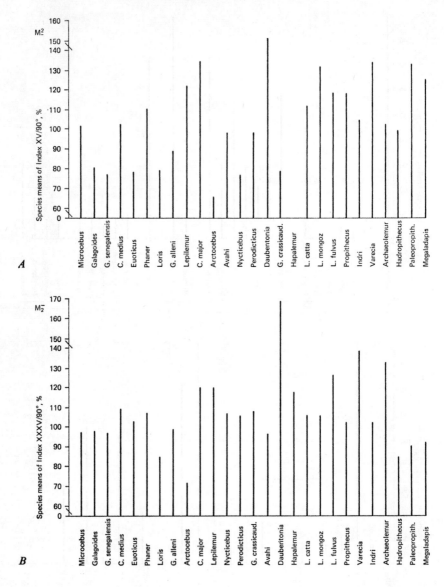

Fig. 35. For legend see page 61.

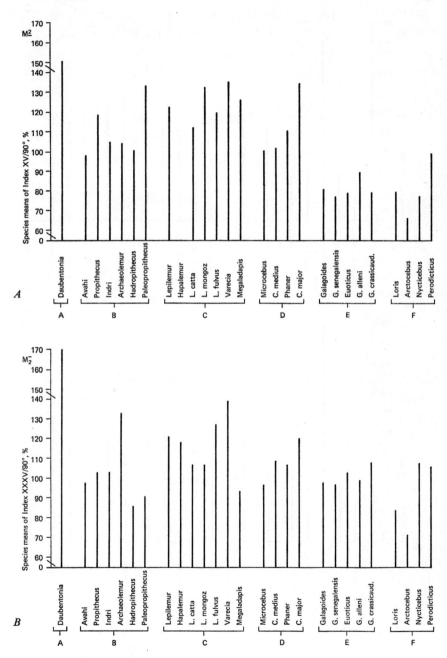

Fig. 36. For legend see page 61.

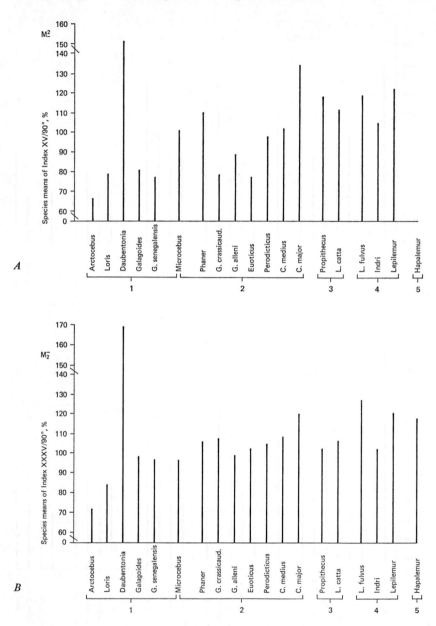

Fig. 37. For legend see page 61.

Legends for figures on pages 56–60.

Fig. 33. Ratio of 'horizontal' to 'vertical' crest curvature in the second molars as it varies with molar size. Compared are the species means for the ratio of Indices (XIII + XXXIII)/(XIV + XXXIV), relating the sum of 'horizontal' crest curvature in the upper and lower molars to the sum of 'vertical' crest curvature in the upper and lower molars. *A* compares species means arranged according to upper molar size (Index II). Molar size increases toward the right. *B* compares species means arranged according to lower molar size (Index XXII). Molar size increases toward the right. Values above 100 indicate a greater emphasis on 'horizontal' crest curvature while values below 100 indicate a greater emphasis on 'vertical' crest curvature.

Fig. 34. Ratio of 'horizontal' to 'vertical' crest curvature in the second molars as it varies taxonomically and with dietary preference. Compared are the species means for the ratio (XIII + XXXIII)/(XIV + XXXIV), relating the sum of 'horizontal' crest curvature in the upper and lower molars to the sum of 'vertical' crest curvature in the upper and lower molars. *A* compares species means grouped by family or subfamily: (A) Daubentoniidae; (B) Indriidae; (C) Lemurinae; (D) Cheirogaleinae; (E) Galaginae; (F) Lorisinae. *B* compares species means arranged according to dietary preference. Table IX explains this arrangement. Values above 100 indicate a greater emphasis on 'horizontal' crest curvature, while values below 100 indicate a greater emphasis on 'vertical' crest curvature.

Fig. 35. Mesio-distal narrowing of the trigon and talonid basins in the second molars as it varies with molar size. *A* compares the M^2 of species with respect to species means of Index XV (angle between preprotocrista and postprotocrista). Species means are arranged according to upper molar size (Index II), with molar size increasing toward the right. *B* compares the M_2 of species with respect to species means of Index XXXV (angle between the cristid obliqua and postcristid). Species means are arranged according to lower molar size (Index XXII), with molar size increasing toward the right. Division of Indices XV and XXXV by 90 degrees was purely arbitrary. Values above 100 indicate that the angle between the appropriate pair of crests is greater than 90 degrees; values below 100 indicate that the angle between the appropriate pair of crests is less than 90 degrees.

Fig. 36. Mesio-distal narrowing of the trigon and talonid basins in the second molars as it varies taxonomically. Species means are grouped by family or subfamily: (A) Daubentoniidae; (B) Indriidae; (C) Lemurinae; (D) Cheirogaleinae; (E) Galaginae; (F) Lorisinae. *A* compares the M^2 of species with respect to species means of Index XV (angle between preprotocrista and postprotocrista). *B* compares the M_2 of species with respect to species means of Index XXXV (angle between cristid obliqua and postcristid). Division of Indices XV and XXV by 90 degrees was purely arbitrary. Values above 100 indicate that the angle between the appropriate pair of crests is greater than 90 degrees. Values below 100 indicate that the angle between the appropriate pair of crests is less than 90 degrees.

Fig. 37. Mesio-distal narrowing of the trigon and talonid basins in the second molars as it varies with dietary preference. Species means are arranged according to dietary preference in a manner explained in table IX. *A* compares the M^2 of species with respect to species means of Index XV (angle between preprotocrista and postprotocrista). *B* compares the M_2 of species with respect to species means of Index XXXV (angle between cristid obliqua and postcristid). Division of Indices XV and XXXV by 90 degrees was purely arbitrary. Values above 100 indicate that the angle between the appropriate pair of crests is greater than 90 degrees. Values below 100 indicate that the angle between the appropriate pair of crests is less than 90 degrees.

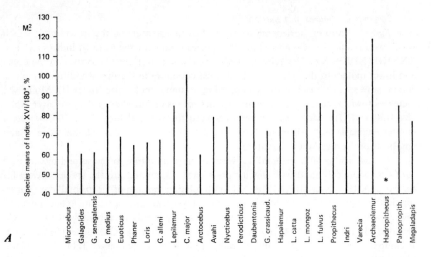

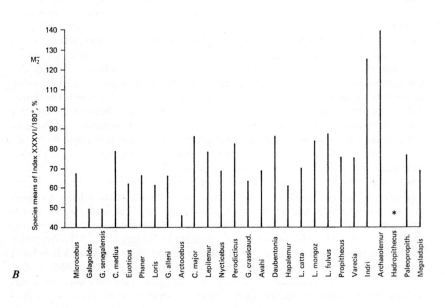

Fig. 38. For legend see page 67.

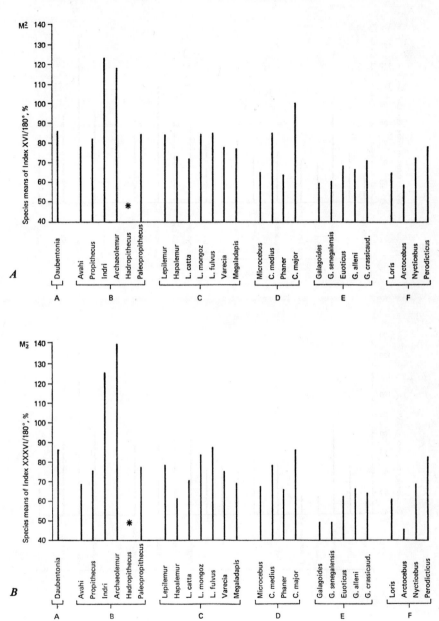

Fig. 39. For legend see page 67.

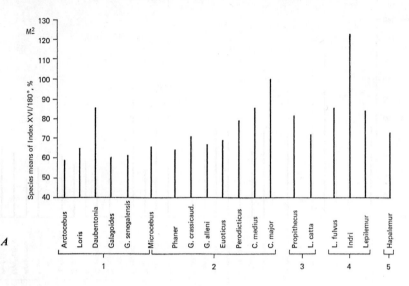

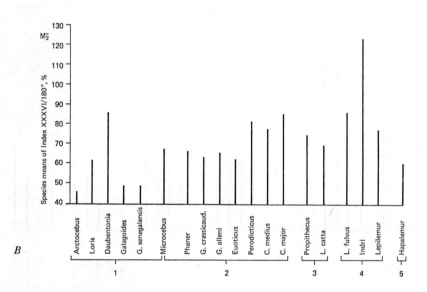

Fig. 40. For legend see page 67.

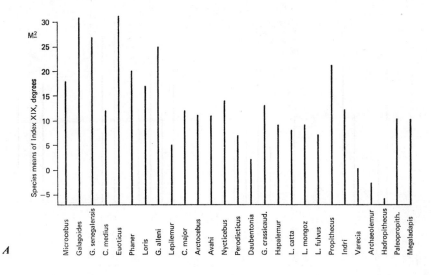

A

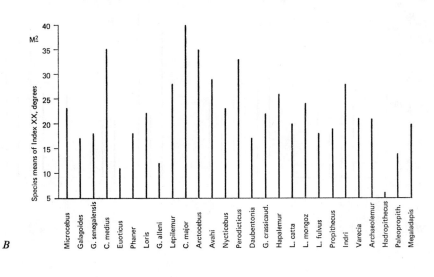

B

Fig. 41. For legend see page 67.

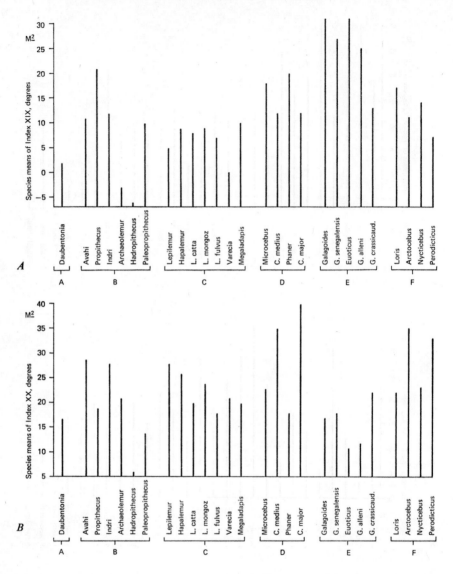

Fig. 42. For legend see page 67.

Legends for figures on pages 62–66

Fig. 38. Mesio-distal contour of the trigon and talonid basins in the second molars as it varies with molar size. *A* compares the M² of species with respect to species means of Index XVI (angle of mesio-distal contour of trigon basin). Species means are arranged according to upper molar size (Index II), with molar size increasing toward the right. *B* compares the M₂̄ of species with respect to species means of Index XXXVI (angle of mesio-distal contour of talonid basin). Species means are arranged according to lower molar size (Index XXII), with molar size increasing toward the right. Division of Indices XVI and XXXVI by 180 degrees was purely for the sake of convenience. The farther values are above 100, the greater is the mesio-distal convexity of the appropriate basin; the farther values are below 100, the greater is the mesio-distal concavity of the appropriate basin. The asterisk above *Hadropithecus* indicates no data available.

Fig. 39. Mesio-distal contour of the trigon and talonid basins in the second molars as it varies taxonomically. Species means are grouped by family or subfamily: (A) Daubentoniidae, (B) Indriidae; (C) Lemurinae; (D) Cheirogaleinae; (E) Galaginae; (F) Lorisinae. *A* compares the M² of species with respect to species means of Index XVI (angle of mesio-distal contour of trigon basin). *B* compares the M₂̄ of species with respect to species means of Index XXXVI (angle of mesio-distal contour of talonid basin). Division of Indices XVI and XXXVI by 180 degrees was purely for the sake of convenience. The farther values are above 100, the greater is the mesio-distal convexity of the appropriate basin; the farther values are below 100, the greater is the mesio-distal concavity of the appropriate basin. The asterisk above *Hadropithecus* indicates no data available.

Fig. 40. Mesio-distal contour of the trigon and talonid basins in the second molars as it varies with dietary preference. Species means are arranged according to dietary preference in a manner explained in table IX. *A* compares the M² of species with respect to species means of Index XVI (angle of mesio-distal contour of trigon basin). *B* compares the M₂̄ of species with respect to species means of Index XXXVI (angle of mesio-distal contour of talonid basin). Division of Indices XVI and XXXVI by 180 degrees was purely for the sake of convenience. The farther values are above 100, the greater is the mesio-distal convexity of the appropriate basin; the farther values are below 100, the greater is the mesio-distal concavity of the appropriate basin.

Fig. 41. Torsion and buccal phase facet orientation in the upper second molars, as they vary with molar size. Species means are arranged according to upper molar size (Index II), with molar size increasing toward the right. *A* compares species means of Index XIX (angle between the cervical plane of the molar crown and the palatal plane). Positive angles indicate molar crown is lingually tilted, negative angles indicate that the molar crown is buccally tilted. *B* compares species means of Index XX (angle between the axis of orientation of buccal phase facetting and the cervical plane of the molar crown). Both Indices XIX and XX represent measurements in the coronal plane.

Fig. 42. Torsion and buccal phase facet orientation in the upper second molars, as they vary taxonomically. Species means are grouped according to family or subfamily: (A) Daubentoniidae;(B)Indriidae;(C)Lemurinae;(D)Cheirogaleinae;(E)Galaginae;(F)Lorisinae. *A* compares species means of Index XIX (angle between the cervical plane of the molar crown and the palatal plane). Positive angles indicate molar crown is lingually tilted, negative angles indicate that the molar crown is buccally tilted. *B* compares species means of Index XX (angle between the axis of orientation of buccal phase facetting and the cervical plane of the molar crown). Both Indices XIX and XX represent measurements in the coronal plane.

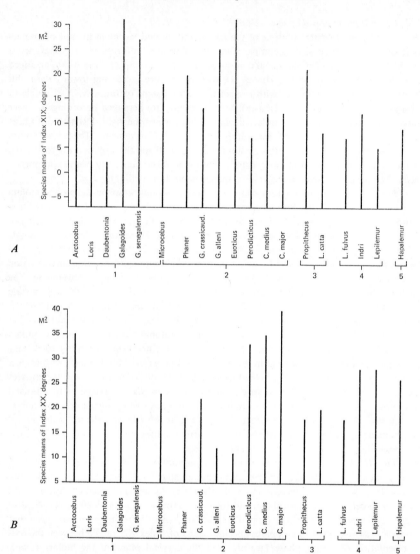

Fig. 43. Torsion and buccal phase facet orientation in the upper second molars, as they vary with dietary preference. Species means are arranged according to dietary preference in a manner explained in table IX. *A* compares species means of Index XIX (angle between the cervical plane of the molar crown and the palatal plane). Positive angles indicate molar crown is lingually tilted, negative angles indicate that the molar crown is buccally tilted. *B* compares species means of Index XX (angle between the axis of orientation of buccal phase facetting and the cervical plane of the molar crown). Both indices XIX and XX represent measurements in the coronal plane.

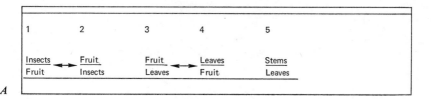

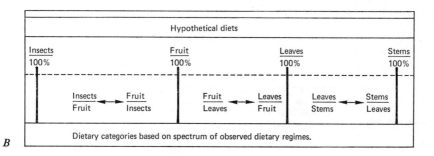

Fig. 44. Utilization and treatment of dietary preferences of the strepsirhines. *A* Sequence of dietary preferences used in histogram analyses of species means of indices of molar form. The arrangement of dietary categories numbered 1–5 provided the framework within which to sequence the known dietary preferences of strepsirhine primates, during histogram analyses. The foods indicated above the line in each dietary category designation represent the primary food preferences, the foods indicated below the line represent the secondary food preferences. Arrows between dietary categories indicate gradation of dietary preferences between the most restrictive dietary regimes with each dietary group. Further explanations are given in 'Dietary Data'. *B* The relationship between hypothetical dietary preferences and actually observed dietary preferences. Hypothetical diets, representing exclusive respective diets of insects, fruits, leaves and stems are indicated at top of vertical bars, above the interrupted line. Dietary categories based on observed dietary regimes are arranged below the interrupted line. Primary food preferences are indicated above the line in each dietary category designation, while secondary food preferences are below the line. Each dietary category, of course, represents a spectrum of dietary preferences, with the most nearly exclusive dietary regimes in each dietary category most closely approaching the appropriate exclusive hypothetical diets. Arrows between dietary categories indicate the gradations of dietary preferences between dietary groups. Further explanations are given in 'Dietary Data'.

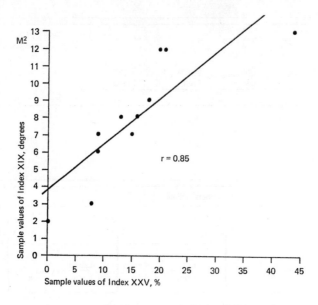

Fig. 45. Bivariate correlation between torsion and wear in the second molars of a sample of individuals of *Lepilemur mustelinus*. The vertical axis represents an angular measurement of the degree of torsion (or buccal eruption) in the upper second molars of *Lepilemur*. The larger the angle (in degrees), the greater is the lingual tilt of the molar crown in the coronal plane. The horizontal axis represents a measurement of dentine exposure and hence wear in the lower second molars of the same sample of individuals of *Lepilemur*. The higher the value (percent), the greater is molar wear. The measurement of wear in the lower second molar was used because it proved to be the most sensitive and reliable indication of overall wear in both upper and lower molars. The slanted solid line represents the regression line for the plotted coordinates; its equation is Y = 26. 73 X + 3.71. The coefficient of correlation (or Pearson's r) for the two variables is given by 'r' and equals 0.85. The regression line and coefficient of correlation indicate that, in *Lepilemur*, as wear in the second molars (ontogenetically) increases, the lingual tilt of the upper second molars also (ontogenetically) increases.

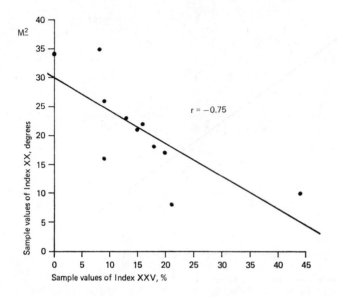

Fig. 46. Bivariate correlation between the orientation of buccal phase facetting and wear in the second molars of a sample of individuals of *Lepilemur mustelinus*. The vertical axis represents an angular measurement of the orientation of buccal phase facetting in the upper second molars of eleven individuals of *Lepilemur*. The larger the angle (in degrees) the more steeply sloped are the buccal phase facets (in the coronal plane) with respect to the cervical plane of the molar crown. The horizontal axis represents a measurement of dentine exposure and hence wear in the lower second molars of the same sample of individuals of *Lepilemur*. The higher the value (percent), the greater is molar wear. (The measurement of wear in the lower second molar was used because it proved to be the most sensitive and reliable indicator of overall wear in both upper and lower molars.) The slanted solid line represents the regression line for the plotted coordinates; its equation is Y = — 57.42 X + 29.94. The coefficient of correlation (or Pearson's r) for the two variables is given by 'r' and equals — 0.75. The regression line and coefficient of correlation indicate that, in *Lepilemur*, as wear in the second molars (ontogenetically) increases, the slope of the buccal phase facets in the upper second molars (ontogenetically) becomes less steep.

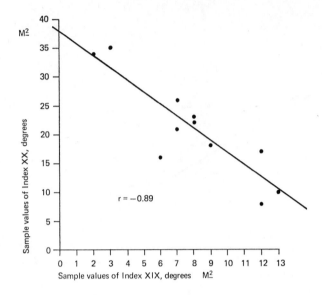

Fig. 47. Bivariate correlation between the orientation of buccal phase facetting and torsion in the upper second molars of a sample of individuals of *Lepilemur mustelinus*. The vertical axis represents an angular measurement of the orientation of buccal phase facetting in the upper second molars of eleven individuals of *Lepilemur*. The larger the angle (in degrees), the more steeply sloped are the buccal phase facets (in the coronal plane) with respect to the cervical plane of the molar crown. The horizontal axis represents an angular measurement of the degree of torsion (or buccal eruption) in the upper second molars of the same sample of individuals of *Lepilemur*. The larger the angle (in degrees), the greater is the lingual tilt of the molar crown in the coronal plane. The slanted solid line represents the regression line for the plotted coordinates; its equation is $Y = -2.15X + 37.95$. The coefficient of correlation (or Pearson's r) for the two variables is given by 'r' and equals -0.89. The regression line and coefficient of correlation indicate that in *Lepilemur*, as the slope of the buccal phase facets in the upper second molars (ontogenetically) becomes less steep, the lingual tilt of the upper second molars (ontogenetically) increases.

IV. Data

A. Dietary Data

Information on the natural dietary regimes of the Strepsirhini was obtained from the most recent and reliable published field observations of each taxon and from personal communications with several workers who have observed various taxa in the wild.

Because few, if any, dietary studies in the wild have been conducted for several consecutive years, and over a large portion of a species geographic range, and because actual weight analyses of the total yearly diet are difficult to undertake, the data on the dietary regimes of many taxa are unfortunately rough estimates.

The natural dietary regimes of several extant strepsirhine species, *Lemur mongoz*, *Varecia variegatus*, *Avahi laniger* and *Nycticebus coucang*, are virtually unknown [SUSSMAN, personal commun.; RICHARD, personal commun.]. For this reason, these species could not be incorporated into analyses relating diet to molar form.

For the purposes of statistical comparison in this investigation, the known dietary regimes of the Strepsirhini were arbitrarily grouped into dietary categories on the basis of estimated primary and secondary food preferences (table IX). A primary dietary preference is here defined as a food accounting for 50% or more of the total yearly dietary intake (by weight) of a species. A secondary dietary preference is here defined as a food accounting for less than 50% of the total yearly dietary intake (by weight) of a species. Dietary categories were established as follows: in category 1, insects are the primary food preference, while fruits and gums are secondarily preferred; in category 2, fruits and gums are primarily preferred, while insects are a secondary food preference; in category 3, fruits and gums are the primary food preferences, and leaves are secondarily preferred; in category 4, leaves are the primary food preference, while fruits and gums are preferred secondarily; in category 5, stems are primarily preferred, and leaves preferred secondarily.

Table IX. Estimated dietary preferences in the strepsirhine primates[1].

Dietary category	Taxon	Sources	Primary %	Secondary %
1	Microcebus murinus	(5,9)	50	50
1	Galago senegalensis	(1,4)	70	30
1	Galagoides demidovii	(2,3)	70	29
1	Daubentonia madagascariensis	(11–13) (14)	75*	25
1	Loris tardigradus	(14)	85	15
1	Arctocebus calabarensis	(2,3)	85	14
2	Phaner furcifer	(11,13,17) (1)	60	40
2	Galago crassicaudatus	(11,13,17) (1)	70	30
2	Galago alleni	(2,3)	73	25
2	Euoticus elegantulus	(2,3)	80	20
2	Perodicticus potto	(2,3)	76	10
2	Cheirogaleus medius	(11,12,17)	80	20
2	Cheirogaleus major	(11,12,17)	85	15
3	Propithecus verreauxi	(8,18)	65	35
3	Lemur catta	(7,8,19,20)	60	40
3	Lemur fulvus	(19,20)	60	40
4	Indri indri	(5,11) (6)	80	20
4	Lepilemur mustelinus	(5,11) (6)	100	00
5	Hapalemur griseus	(10,15,16)	70	30

1 Estimated dietary preferences in the strepsirhine primates broken down by dietary categories: 1) insects and small vertebrates primarily preferred, fruits and gums secondarily preferred; 2) fruits and gums primarily preferred, insects and small vertebrates secondarily preferred; 3) fruits and gums primarily preferred, leaves secondarily preferred; 4) leaves primarily preferred, fruits and gums secondarily preferred; 5) stems primarily preferred, leaves secondarily preferred. Below each taxon listed are given the estimated percentages of the total diet (by weight) accounted for by the primary and secondary food preferences (above and below the line, respectively). The asterisk next to the primary food preference of Daubentonia indicates that the insect component of the diet of this taxon consists mainly of soft insect larvae. The numbers in parentheses above each taxon listed refer to the sources of the dietary data for each taxon: (1) BEARDER and DOYLE, 1974; (2) CHARLES-DOMINIQUE, 1971; (3) CHARLES-DOMINIQUE, 1974; (4) DOYLE, 1974; (5) HLADICK, personal commun.; (6) HLADICK and CHARLES-DOMINIQUE, 1974; (7) JOLLY, 1966; (8) JOLLY, 1967; (9) MARTIN, 1973; (10) MILTON, personal commun.; (11) PETTER, 1962a; (12) PETTER, 1962b; (13) PETTER, 1965; (14) PETTER and HLADICK, 1970; (15) PETTER and PEYRIERAS, 1970; (16) PETTER and PEYRIERAS, 1975; (17) PETTER et al., 1971; (18) RICHARD, personal commun.; (19) SUSSMAN, 1974; (20) SUSSMAN, personal commun.

The several models of molar adaptation which I have developed assume nearly exclusive respective diets of insects, fruits, leaves and stems. These hypothetical diets thus represent dietary 'extremes' within the broad spectrum of strepsirhine dietary regimes observed in the wild and are diagrammatically related to the dietary categories based on actual field observations in figure 44 B.

While the concept of 'dietary category' is a convenient one in terms of generating and testing heuristic models and grouping data, it is realized that in an evolutionary sense it is quite inappropriate. The dietary niche of each species is, in a sense, unique, while the spectrum of possible dietary regimes which can be adopted by an evolving species is virtually infinite.

B. Figures

Line drawings were made by the author of the left M^2 and the right $M_{\overline{2}}$ from representative species of extant and subfossil strepsirhines (fig. 48–68).

The line drawings of molars from extant species were made from dried skulls with the aid of a Wild M-5 stereoscope and a camera lucida. Four drawings were made of both the M^2 and $M_{\overline{2}}$. For both molars, line drawings were made of the occlusal, disto-bucco-occlusal and mesio-linguo-occlusal views. In addition to these drawings, a disto-bucco-occlusal view provided the basis for a 'contour line' drawing of the trigon basin surfaces of the M^2, while a mesio-linguo-occlusal view provided the basis for 'contour line' drawing of the trigon basin surfaces of the $M_{\overline{2}}$. Where possible, patterns of buccal phase facetting were indicated by hatched areas along molar crests. The hatching was drawn parallel to the striations seen on these facets. Occasionally, lingual phase wear surfaces were recognizably striated, and these were indicated in the drawings of a few species.

Line drawings of the molars of subfossil strepsirhines were made by tracing over the projected images of color slides of these teeth. Only occlusal views were drawn for the M^2 and $M_{\overline{2}}$.

The line drawings not only provide an eloquent documentation of the molar morphology of each species studied, but also arouse an awareness of the great interspecific variety of form demonstrated by homologous molar features. Drawings such as these also serve as a sobering reminder that morphology cannot be totally quantified, and that aspects of form are lost in the translation into numbers.

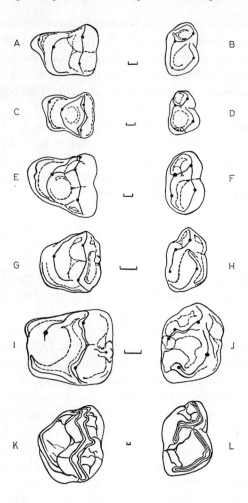

Fig. 48. Occlusal views of the second molars of the Lemurinae. *A, C, E, G, I, K,* upper molars; *B, D, F, H, J, L,* lower molars. *A* and *B, Lemur fulvus* (A.M.N.H. 77828); *C* and *D, Lemur catta* (A.M.N.H. 170739); *E* and *F, Varecia variegatus* (A.M.N.H. 77828); *G* and *H, Lepilemur mustelinus* (A.M.N.H. 170576); *I* and *J, Hapalemur griseus* (A.M.N.H. 170672); *K* and *L, Megaladapis edwardsi.* Mesial is toward the top, buccal is to the right. Each scale represents approximately 1 mm.

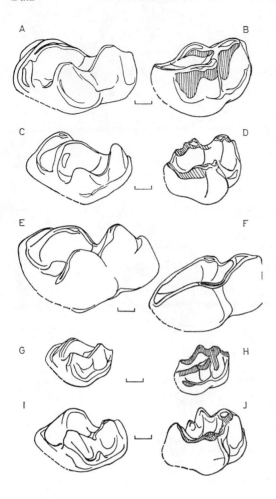

Fig. 49. Disto-bucco-occlusal views of the second molars of the Lemurinae. *A, C, E, G, I,* upper molars; *B, D, E, H, J,* lower molars. *A* and *B, Lemur fulvus* (A.M.N.H. 170723); *C* and *D, Lemur catta* (A.M.N.H. 170739); *E* and *F, Varecia variegatus* (A.M. N.H. 77828); *G* and *H, Lepilemur mustelinus* (A.M.N.H. 170576); *I* and *J, Hapalemur griseus* (A.M.N.H. 170672). Each scale represents approximately 1 mm.

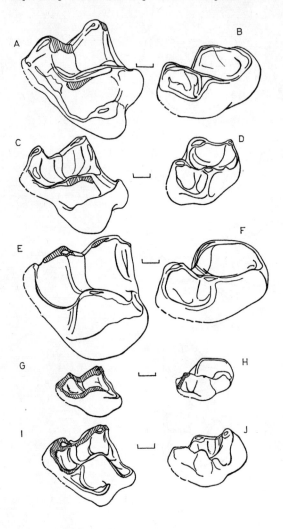

Fig. 50. Mesio-linguo-occlusal views of the second molars of the Lemurinae.
A, C, E, G, I, upper molars; *B, D, F, H, J*, lower molars. *A* and *B* ,*Lemur fulvus* (A.M.
N.H. 170723); *C* and *D, Lemur catta* (A.M.N.H. 170739); *E* and *F* , *Varecia variegatus*
(A.M.N.H. 77828); *G* and *H, Lepilemur mustelinus* (A.M.N.H. 170576); *I* and *J, Hapa-*
lemur griseus (A.M.N.H. 170672). Each scale represents approximately 1 mm.

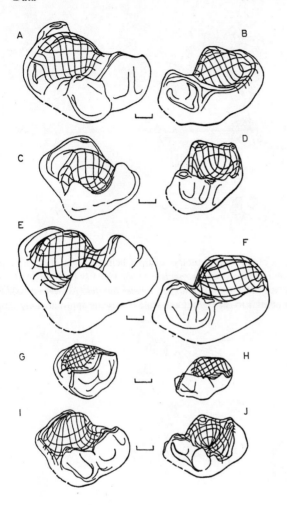

Fig. 51. Contours of the trigon and talonid basins of the second molars of the Lemurinae. *A, C, E, G, I,* disto-bucco-occlusal views of upper molars; *B, D, F, H, J,* mesio-linguo-occlusal views of the lower molars. *A* and *B, Lemur fulvus* (A.M.N.H. 170723); *C* and *D, Lemur catta* (A.M.N.H. 170739); *E* and *F, Varecia variegatus* (A.M.N.H. 77828); *G* and *H, Lepilemur mustelinus* (A.M.N.H. 170576); *I* and *J, Hapalemur griseus* (A.M.N.H. 170672). Each scale represents approximately 1 mm.

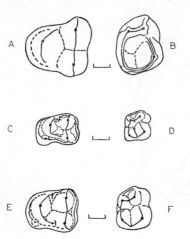

Fig. 52. Occlusal views of the second molars of the Cheirogaleinae. *A, C, E,* upper molars; *B, D, F,* lower molars. *A* and *B, Cheirogaleus major* (A.M.N.H. 100640); *C* and *D, Microcebus murinus* (A.M.N.H. 100804); *E* and *F, Phaner furcifer* (A.M.N.H. 100829). Mesial is toward the top, buccal is to the right. Each scale represents approximately 1 mm.

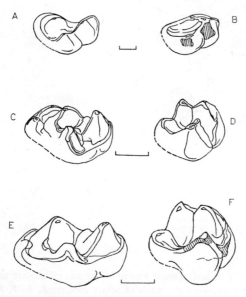

Fig. 53. Disto-bucco-occlusal views of the second molars of the Cheirogaleinae. *A, C, E,* upper molars; *B, D, F,* lower molars. *A* and *B, Cheirogaleus major* (A.M.N.H. 100640); *C* and *D ,Microcebus murinus* (A.M.N.H. 100804); *E* and *F, Phaner furcifer* (A.M.N.H. 100829). Each scale represents approximately 1 mm.

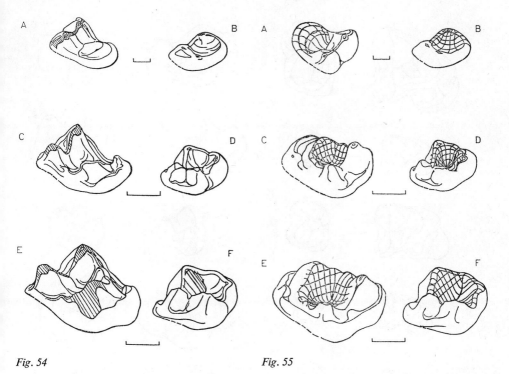

Fig. 54   Fig. 55

Fig. 54. Mesio-linguo-occlusal views of the second molars of the Cheirogaleinae.
A, C, E, upper molars; B, D, F, lower molars. A and B, Cheirogaleus major (A.M.N.H.
100640); C and D, Microcebus murinus (A.M.N.H. 100804); E and F, Phaner furcifer
(A.M.N.H. 100829). Each scale represents approximately 1 mm.

Fig. 55. Contours of the trigon and talonid basins of the second molars of the
Cheirogaleinae. A, C, E, disto-bucco-occlusal views of upper molars; B, D, E, mesio-linguo-
occlusal views of lower molars. A and B, Cheirogaleus major (A.M.N.H. 100640); C and
D, Microcebus murinus (A.M.N.H. 100804); E and F, Phaner furcifer (A.M.N.H. 100829).
Each scale represents approximately 1 mm.

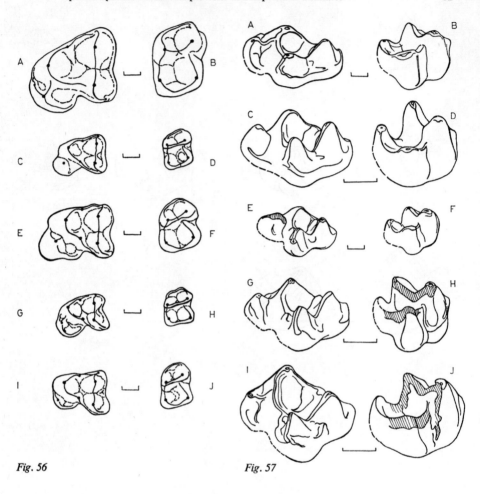

Fig. 56   Fig. 57

Fig. 56. Occlusal views of the second molars of the Galaginae. A, C, E, G, I, upper molars; B, D, E, H, J, lower molars. A and B, Galago crassicaudatus (A.M.N.H. 116248); C and D, Galago senegalensis (A.M.N.H. 81458); E and F, Galago alleni (A.M.N.H. 241120); G and H, Galagoides demidovii (A.M.N.H. 50975); I and J, Euoticus elegantulus (A.M.N.H. 241126). Mesial is toward the top, buccal is to the right. Each scale represents approximately 1 mm.

Fig. 57. Disto-bucco-occlusal views of the second molars of the Galaginae. A, C, E, G, I, upper molars; B, D, F, H, J, lower molars. A and B, Galago crassicaudatus (A.M. N.H. 116248); C and D, Galago senegalensis (A.M.N.H. 81458); E and F, Galago alleni (A.M.N.H. 24120); G and H, Galagoides demidovii (A.M.N.H. 50975); I and J, Euoticus elegantulus (A.M.N.H. 241126). Each scale respresents approximately 1 mm.

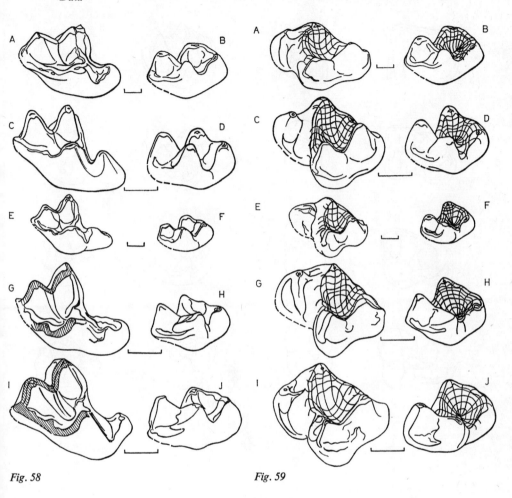

Fig. 58

Fig. 59

Fig. 58. Mesio-linguo-occlusal views of the second molars of the Galaginae. *A, C, E, G, I,* upper molars; *B, D, F, H, I,* lower molars. *A* and *B, Galago crassicaudatus* (A.M. N.H. 116248); *C* and *D, Galago senegalensis* (A.M.N.H. 81458); *E* and *F, Galago alleni* (A.N.M.H. 24120); *G* and *H, Galagoides demidovii* (A.M.N.H. 50975); *I* and *J, Euoticus elegantulus* (A.M.N.H. 241126). Each scale represents approximately 1 mm.

Fig. 59. Contours of the trigon and talonid basins of the second molars of the Galaginae. *A, C, E, G, I,* disto-bucco-occlusal views of upper molars; *B, D, F, H, J,* mesio-linguo-occlusal views of the lower molars. *A* and *B, Galago crassicaudatus* (A.M.N.H. 116248); *C* and *D, Galago senegalensis* (A.M.N.H. 81458); *E* and *F, Galago alleni* (A.N.M.H. 24120); *G* and *H, Galagoides demidovii* (A.M.N.H. 50975); *I* and *J, Euoticus elegantulus* (A.M.N.H. 241126). Each scale represents approximately 1 mm.

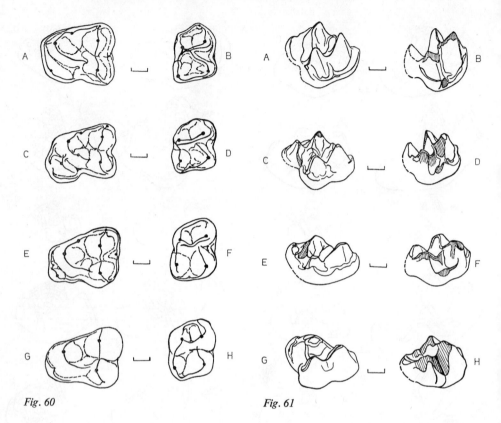

Fig. 60

Fig. 61

Fig. 60. Occlusal views of the second molars of the Lorisinae. *A, C, E, G,* upper molars; *B, D, E, H,* lower molars. *A* and *B, Arctocebus calabarensis* (A.M.N.H. 207949); *C* and *D, Loris tardigradus* (A.M.N.H. 217303); *E* and *F, Nycticebus coucang* (A.M.N.H. 102845); *G* and *H, Perodicticus potto* (A.M.N.H. 52688). Mesial is toward the top, buccal is to the right. Each scale represents approximately 1 mm.

Fig. 61. Disto-bucco-occlusal views of the second molars of the Lorisinae. *A, C, E, G,* upper molars; *B, D, F, H,* lower molars. *A* and *B, Arctocebus calabarensis* (A.M. N.H. 207949); *C* and *D, Loris tardigradus* (A.M.N.H. 217303); *E* and *F, Nycticebus coucang* (A.M.N.H. 102845); *G* and *H, Perodicticus potto* (A.M.N.H. 52688). Each scale represents approximately 1 mm.

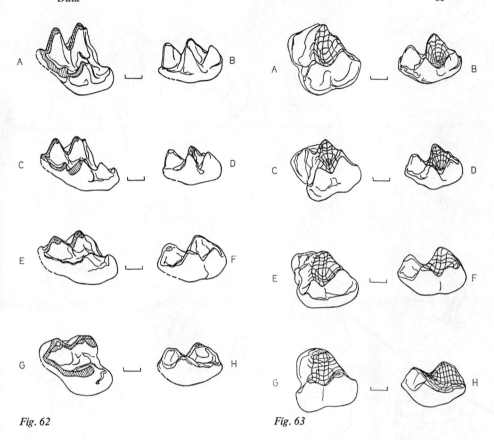

Fig. 62

Fig. 63

Fig. 62. Mesio-linguo-occlusal views of the second molars of the Lorisinae. *A, C, E, G,* upper molars; *B, D, F, H,* lower molars. *A* and *B, Arctocebus calabarensis* (A.M. N.H. 207949); *C* and *D, Loris tardigradus* (A.M.N.H. 217303); *E* and *F, Nycticebus coucang* (A.M.N.H. 102845); *G* and *H, Perodicticus potto* (A.M.N.H. 52688). Each scale represents approximately 1 mm.

Fig. 63. Contours of the trigon and talonid basins of the second molars of the Lorisinae. *A, C, E, G,* disto-bucco-occlusal views of upper molars; *B, D, F, H,* mesio-linguo-occlusal views of the lower molars. *A* and *B, Arctocebus calabarensis* (A.M.N.H. 207949); *C* and *D, Loris tardigradus* (A.M.N.H. 217303); *E* and *F, Nycticebus coucang* (A.M.N.H. 102845); *G* and *H, Perodicticus potto* (A.M.N.H. 52688). Each scale represents approximately 1 mm.

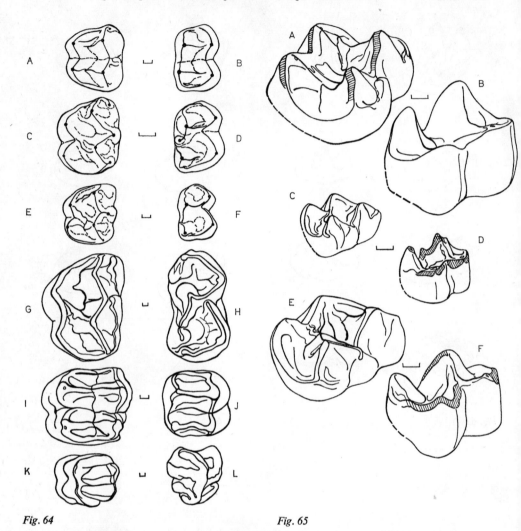

Fig. 64 *Fig. 65*

Fig. 64. Occlusal views of the second molars of the Indriidae. *A, C, E, G, I, K,* upper molars; *B, D, F, H, J, L,* lower molars. *A* and *B, Indri indri* (A.M.N.H. 100503); *C* and *D, Avahi laniger* (A.M.N.H. 41267); *E* and *F, Propithecus verreauxi* (A.M.N.H. 16699). *G* and *H, Palaeopropithecus ingens; I* and *J, Archaeolemur majori; K* and *L, Hadropithecus stenognathus.* Mesial is toward the top, buccal is to the right. Each scale represents approximately 1 mm.

Fig. 65. Disto-bucco-occlusal views of the second molars of the Indriidae. *A, C, E,* upper molars; *B, D, F,* lower molars; *A* and *B, Indri indri* (A.M.N.H. 100504); *C* and *D, Avahi laniger* (A.M.N.H. 41267); *E* and *F, Propithecus verreauxi* (A.M.N.H. 16699). Each scale represents approximately 1 mm.

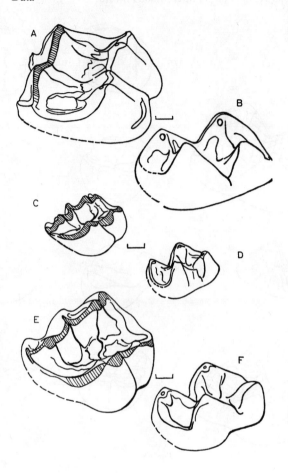

Fig. 66. Mesio-linguo-occlusal views of the second molars of the Indriidae. *A, C, E*, upper molars; *B, D, F*, lower molars. *A* and *B, Indri indri* (A.M.N.H. 100504); *C* and *D, Avahi laniger* (A.M. N.H. 41267); *E* and *F, Propithecus verreauxi* (A.M.N.H. 16699). Each scale represents approximately 1 mm.

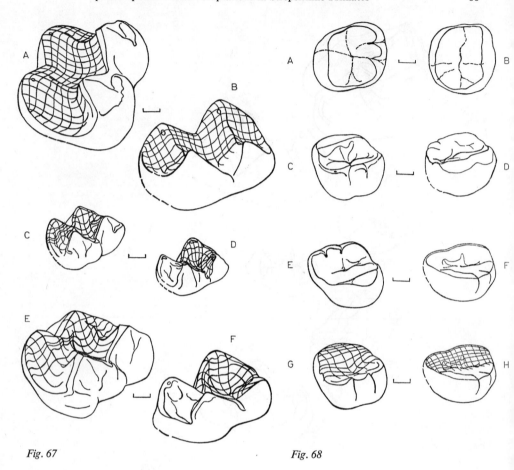

Fig. 67.   Fig. 68.

Fig. 67. Contours of the trigon and talonid basins of the second molars of the Indriidae. *A, C, E,* disto-bucco-occlusal views of upper molars; *B, D, E,* mesio-linguo-occlusal views of lower molars. *A* and *B, Indri indri* (A.M.N.H. 100504); *C* and *D, Avahi laniger* (A.M.N.H. 41267); *E* and *F, Propithecus verreauxi* (A.M.N.H. 16669). Each cale represents approximately 1 mm.

Fig. 68. Various views of the second molars of *Daubentonia madagascariensis* (A.M. N.H. 100632). *A, C, E, G,* upper molars; *B, D, E, H,* lower molars. *A* and *B,* occlusal views, with mesial toward the top and buccal to the right; *C* and *D,* disto-bucco-occlusal views; *E* and *F,* mesio-linguo-occlusal views; *G,* disto-bucco-occlusal view of upper molar, showing contour of the trigon basin; *H,* mesio-linguo-occlusal view of lower molar, showing contour of talonid basin. Each scale represents approximately 1 mm.

V. A Survey of the Morphology of
Upper and Lower Second Molars in Strepsirhines

In the following section, the inferred ancestral morphotypes for the molars of each family or subfamily will be outlined, while diagnoses will be presented of the molars of each species. Illustrations will complement this presentation.

Reconstructions of ancestral morphotypes were arrived at by abstracting, as best as possible, those molar features which were considered primitive for a group of species related at the family or subfamily level. This process involved estimating which molar features were held by most species within a family or subfamily, and then specifying which extant species most nearly embodied these features.

Species diagnoses contrasted species within the same family or subfamily with respect to presumably derived functional morphological features. Those concepts and terms which relate to functional molar morphology and which were used in these diagnoses are explained in 'Hypotheses'.

Figures 48–68 each group comparable views of the upper and lower second molars of species within the same family or subfamily, and may be referred to in the course of this presentation. Explanations of the drawings in these figures are given in the figure legends and in 'Data'.

A. Possible Ancestral Morphotypes (and Attendant Heritage Features) of the M^2 and $M_{\overline{2}}$ of Each Family or Subfamily

1. Lemurinae

The Lemurinae are a morphologically rather diverse group for which an ancestral morphotype is not easy to reconstruct. The genera, *Lemur* and *Varecia*, contrast in many features with *Lepilemur*, *Megaladapis* and *Hapalemur*. It is suggested here, after weighing all the morphological evidence, that the ancestral lemurine may have possessed an upper second molar similar to that of *Lepilemur*, and a lower second molar most similar to that of *Megaladapis*. The ancestral lemurine M^2 would probably have differed, however, from that of *Lepilemur* in possessing a more in-flated ectoloph, and a mesio-distally narrower protocone, while the an-cestral $M_{\overline{2}}$ would have differed from that of *Megaladapis* in lacking a metaconulid, and perhaps demonstrating a long, continuous, but low relief entocristid.

2. Cheirogaleinae

The Cheirogaleinae are morphologically an even more diverse group than the Lemurinae, and efforts to reconstruct their ancestral morphotype must be most speculative at best. The second molars of all cheirogaleines demonstrate similar arrangements of cusps and crests, and all show cusp inflation to varying degrees. It is suggested that the upper and lower second molars of the ancestral cheirogaleine may have been most similar to those of *Cheirogaleus*, but may have differed from the latter taxa in demonstrating somewhat greater cusp and crest relief.

3. Galaginae

The galagines, while displaying some intergeneric differences, appear to demonstrate relatively great morphological uniformity. The M^2 and $M_{\overline{2}}$ of *Galago alleni*, as they most nearly embody those molar features most commonly held in this subfamily, perhaps represent as good approximations as any, to the second molars of the probable ancestral galagine.

4. Lorisinae

The genera within the Lorisinae display a fair amount of morphological diversity. This fact renders any ancestral reconstructions for this group more difficult than for the Galaginae. It would appear that the M^2 and $M_{\overline{2}}$ of *Nycticebus* most nearly embody those molar features possessed by most lori-sines, and as such, would appear to most closely resemble the second molars of the ancestral lorisine. The ancestral lorisine may have demonstrated some-what greater cusp relief.

5. Indriidae

If it is assumed that the variably crosslophed (or bilophed) indriid molars ultimately derive from an ancestor which lacked molar crosslophs, it may be reasonable to suggest that the second molars of the most recent common indriid ancestor most closely resembled those of *Palaeopropithecus*. In terms of molar form, *Archaeolemur and Hadropithecus* would appear to be the most derived genera.

6. Daubentoniidae

This taxon is represented by only one genus, which displays virtually no interspecific variability in molar form. Ancestral molar features for this family, therefore, cannot be inferred.

B. Species Diagnoses (Highlighting Interspecific Differences in Presumably Derived Functional Morphological Features) of the Upper and Lower Second Molars

1. Lemur fulvus (and Lemur macaco) (fig. 48 A + B, 49A + B, 50A + B, 51A + B)

The M^2 and $M_{\overline{2}}$ of *Lemur fulvus* and *Lemur macaco* differ from those of *Lemur catta* in possessing more inflated cusps of lower relief and acuity, while their protocones and hypoconids, respectively, are mesio-distally very broad, with very extensive, planar incusion surfaces. The trigon and talonid basins are far more shallow and mesio-distally virtually unconfined, while cresting tends to be more mesio-distally oriented.

These features indicate a greatly reduced functional emphasis on point penetration, and greater emphases on 'horizontal' point-cutting and efficient crushing and grinding.

2. Lemur catta (fig. 48 C + D, 49 C + D, 50 C + D, 51 C + D)

The second molars of *Lemur catta* differ from those of other congeneric species in respectively demonstrating more acute cusps and relatively deep, and smoothly concave trigon and talonid basins which are fully confined by relatively high relief, and greatly narrowed, crests. The protocone and hypoconid are less wide mesio-distally. The $M_{\overline{2}}$ is particularly distinctive in emphasizing relatively great relief in the protocristid, postcristid and entoconid.

These features would appear to indicate relatively greater functional emphases on point penetration, 'punching' and 'vertical' point cutting.

3. Varecia variegatus (fig. 48 E + F, 49 E + F, 50 E + F, 51 E + F)

The M^2 of *V. variegatus* differs from that of *Lemur fulvus* in demonstrating more acute cusps and a much less extensive protocone which is rotated so as to mesially confine and distally open the trigon basin. The $M_{\overline{2}}$ is distinguished from the latter species in demonstrating a deep, mesio-distally oriented, trough-like talonid basin, which opens distally and which is well confined buccally and lingually by crests of higher relief.

These features are of uncertain functional significance, but may suggest greater emphases on puncturing and bending.

4. *Lepilemur mustelinus* (fig. 48 G + H, 49 G + H, 50 G + H, 51 G + H)

The second molars of *Lepilemur* greatly contrast with those of *Hapalemur* in demonstrating squat, mesio-distally elongated cusps of reduced relief and acuity, mesio-distally extensive and planar incusion surfaces in the trigon and talonid basins, respectively, and very sharp-edged crests which emphasize a mesio-distal orientation.

These features suggest overwhelming functional emphases on 'horizontal' point-cutting and efficient crushing and grinding.

5. *Hapalemur griseus* (fig. 48 I + J, 49 I + J, 50 I + J, 51 I + J)

The second molars of *Hapalemur* contrast sharply with those of *Lepilemur* in demonstrating higher relief, more acute and more conical cusps (especially the lingual ones). The M^2 and $M_{\overline{3}}$ also differ as they respectively demonstrate basically deep, concave trigon and talonid basins, which are mostly well confined, except for large, localized gaps in the basin walls. The $M_{\overline{3}}$ most notably emphasizes more transversely oriented crests.

These features suggest great functional emphases on puncturing, bending, and 'punching', and a relatively greater emphasis on 'vertical' point-cutting.

6. *Megaladapis edwardsi* (fig. 48 K + L)

The M^2 and $M_{\overline{3}}$ of *Megaladapis* are very similar morphologically, and broadly similar functionally, to those of *Lepilemur*. The M^2 of the former differs from that of the latter in possessing a more inflated ectoloph, while the $M_{\overline{3}}$ of *Megaladapis*, unlike that of *Lepilemur*, lacks an entocristid but demonstrates the presence of a metaconulid and a mesio-distally narrower hypoconid.

7. *Cheirogaleus major* (fig. 52 A + B, 53 A + B, 54 A + B, 55 A + B)

The second molars of *Cheirogaleus major* greatly differ from those of *Phaner*, and especially from those of *Microcebus* in demonstrating greater mesio-distal width, very blunt, squat cusps of exceedingly low relief, crests of minimal length, salience and sharpness, and basins which are virtually undefined.

These features all suggest greatly reduced functional emphases on point penetration and point-cutting, and instead suggest capabilities limited almost totally to crushing and grinding.

8. Microcebus murinus (fig. 52 C + D, 53 C + D, 54 C + D, 55 C + D)

The M^2 and $M_{\overline{2}}$ differ greatly from those of *Cheirogaleus* in demonstrating great mesio-distal constriction, more conical cusps of relatively great relief and acuity, respective trigon and talonid basins of great depth, concavity and mesio-distal confinement, and relatively long, salient, sharp-edged crests.

These features suggest much greater functional emphases on point penetration and 'vertical' point-cutting.

9. Phaner furcifer (fig. 52 E + F, 53 E + F, 54 E + F, 55 E + F)

The second molars of *Phaner* differ from those of *Microcebus* in being mesio-distally slightly wider, with slightly reduced cusps relief and acuity. Molars also demonstrate a great reduction in relative size.

The functional capabilities of the M^2 and $M_{\overline{2}}$ of *Phaner* are similar to, but somewhat more reduced than, those of *Microcebus*.

10. Galago (fig. 56–59)

The M^2 and $M_{\overline{2}}$ of the species included in the genus, *Galago*, roughly constitute a morphological cline, with those of *G. crassicaudatus* and *G. senegalensis* at the two extremes. The second molars of *G. alleni* (fig. 56 E + F, 57 E + F, 58 E + F, 59 E + F), though in some ways intermediate, are basically very similar to those of *G. crassicaudatus*. Only the M^2 and $M_{\overline{2}}$ of the more 'extreme' species will be contrasted, with the nature of the second molars of *G. alleni* assumed as understood.

11. Galago crassicaudatus (fig. 56 A + B, 57 A + B, 58 A + B, 59 A + B)

The M^2 and $M_{\overline{2}}$ of *G. crassicaudatus*, when compared to those of *G. senegalensis*, demonstrate greatly reduced cusp relief and acuity, greatly rounded crest edges, greatly reduced basin depth, concavity, and confinement, and greater mesio-distal width.

These features appear to suggest that point penetration and 'vertical' point-cutting are greatly de-emphasized functions, while low penetrative, incusive functions are emphasized.

12. Galago senegalensis (fig. 56 C + D, 57 C + D, 58 C + D, 59 C + D)

The second molars of *G. senegalensis*, by contrast, demonstrate great mesio-distal constriction, conical cusps of very great relief and acuity, basins of greath depth, concavity and mesio-distal confinement, and a marked emphasis on transversely oriented, sharp-edged crests which demonstrate greater relief.

These features suggest that efficient point penetration and 'vertical' point-cutting are very strongly emphasized functions.

13. *Galagoides demidovii* (fig. 56G + H, 57G + H, 58G + H, 59G + H)

The M^2 and $M_{\overline{3}}$ of *G. demidovii* are very similar to those of *G. senegalensis*, both morphologically and functionally, except, most notably, for the fact that the molars of the former species demonstrate crests of somewhat greater sharpness and angularity.

14. *Euoticus elegantulus* (fig. 56I + J, 57I + J, 58I + J, 59I + J)

The second molars of this species are similar to those of *G. demidovii*, but differ from those of the latter in being more mesio-distally elongated. They also demonstrate cusps of slightly lower relief and acuity, basins of somewhat less depth and concavity, and crests of less transverse orientation. The molars also demonstrate a great reduction in relative size.

These features suggest the somewhat reduced functional emphases on point penetration and 'vertical' point-cutting.

15. *Arctocebus calabarensis* (fig. 60A + B, 61A + B, 62A + B, 63A + B)

The M^2 and $M_{\overline{3}}$ of *Arctocebus* contrast with those of all other lorisines in demonstrating greatly compressed (i.e., narrow) crests. The molars of this species also exceed those of all other lorisines (including *Loris*) in cusp relief and acuity. Compared to the molars of *Perodicticus* and *Nycticebus*, those of *Arctocebus* demonstrate very conical cusps as well as crests of great length, salience and sharpness which emphasize a transverse orientation. Basins are very deep, greatly concave and mesio-distally greatly confined.

These features suggest very strong functional emphases on efficient point penetration and 'vertical' point-cutting.

16. *Loris tardigradus* (fig. 60C + D, 61C + D, 62C + D, 63C + D)

The second molars of *Loris* are basically similar to those of *Arctocebus*, both morphologically and functionally, but differ from those of the latter in demonstrating somewhat less cusp relief and acuity, and somewhat more cusp and crest inflation.

17. *Perodicticus potto* (fig. 60G + H, 61G + H, 62G + H, 63G + G)

The M^2 and $M_{\overline{3}}$ of *Perodicticus* strongly contrast with those of *Arctocebus* and *Loris* in demonstrating blunt, very low relief and greatly inflated cusps, as well as very shallow basins which are very weakly concave and very poorly confined. Crests are shortened and blunt, and do not rigidly adhere to a transverse orientation. Molars also demonstrate moderately reduced relative size.

These features indicate a great reduction in the functional efficiency of point penetration and 'vertical' point-cutting, and suggest instead emphases on incusive functions.

18. Nycticebus coucang (fig. 60 E + F, 61 E + F, 62 E + F, 63 E + F)

The second molars of this species are very similar to those of *Perodicticus*, but differ from those of the latter in demonstrating somewhat greater cusp and crest relief, and basin depth.

These features perhaps suggest that the molars of *Nycticebus* somewhat more efficiently perform point penetration and 'vertical' point-cutting than do those of *Perodicticus*.

19. Palaeopropithecinae and Indriinae (fig. 64–67)

The M^2 and $M_{\overline{2}}$ of *Palaeopropithecus ingens*, *Propithecus verreauxi*, *Avahi laniger* and *Indri indri* form a morphological cline, with molar features associated with crosslophing least evident in *Palaeopropithecus*, and most evident in *Indri*. The transverse alignment and differential relief of the buccal and lingual cusps, the parity in size of the mesial and distal moieties of the molars, the development of crosslophs and the obliteration of the primitive basins, and the shift in the termination of the protocristid from the apex of the metaconid to the centroflexid are all least evident in *Palaeopropithecus*, somewhat more evident in *Propithecus*, still more demonstrated in *Avahi*, and most greatly demonstrated in *Indri*.

These trends in molar features suggest that the functional capacities to both extensively point-cut and pleat food are greatest in *Indri*, less in *Avahi* and *Propithecus*, and least by far in *Palaeopropithecus*.

20. Archaeolemur majori (fig. 64 I + J)

The second molars of *Archaeolemur* differ from those of the indriines by demonstrating more mesio-distal compression, while their far more prominent crosslophs virtually absorb all cusps. Crest sharpness is, however, more reduced.

Though highly crosslophed, the molars of *Archaeolemur* apparently de-emphasize point-cutting.

21. Hadropithecus stenognathus (fig. 64 K + L)

The M^2 and $M_{\overline{2}}$ of this species are roughly similar to those of *Archaeo-lemur*, but differ in possessing sharper crests and mesio-distally much more compressed crosslophs, which are surrounded mesially and distally by, and confluent lingually (for M^2) or buccally (for $M_{\overline{2}}$) with, thick, continuous ridges which border the molar crowns.

The truncation of crown features with wear generates long, greatly curved and horizontal cutting edges. These edges perform extensive point-cutting.

22. Daubentonia madagascariensis (fig. 68)

The second molars of this species differ from those of the Indriidae in demonstrating nearly flat and featureless quadrate crowns which are surrounded by smoothly rounded borders.

Molar function is apparently severely limited to crushing and grinding.

VI. Discussion

The salient patterns of strepsirhine molar adaptations which emerge from the illustrative-descriptive and analyzed metrical data are discussed in the following section. The discussion is broken down into units, each of which deals with a class of related features of molar form, wear or orientation, and which invokes all relevant non-metrical and metrical data. The vast bulk of the discussion is devoted to species-specific molar adaptations, while a small portion treats ontogenetic patterns of molar wear and molar reorientation.

A. Species-Specific Molar Adaptations

1. Cusp Relief and Acuity

As figures 48–68 and the species diagnoses testify, cusp relief, form and acuity vary greatly among strepsirhines. Histogram analyses of Indices VI (fig. 15A, 16A, 17A) and XXVI (fig. 15B, 16B, 17B); VII (fig. 18A, 19A, 20A) and XXVII (fig. 18B, 19B, 20B); and VIII (fig. 21A, 22A, 23A) and XVIII (fig. 21B, 22B, 23B) reveal the patterns of interspecific variation in cusp relief and acuity related to molar size (fig. 15, 18, 21), heritage (fig. 16, 19, 22), and dietary preference (fig. 17, 20, 23). In each pair of comparable indices, the values for M^2-demonstrate an especially strong separation according to dietary preference. The values for $M_{\overline{3}}$ demonstrate stronger size- and heritage-related variation than do the M^2 values.

The size-related variation for the $M_{\overline{3}}$ values appears to be especially strong, though it does demonstrate a relatively great deal of fluctuation within the lower range of molar sizes. The $M_{\overline{3}}$ values also separate out according to dietary preference in a manner similar to, but more subdued than, the values for the M^2.

The relatively strong diet-related variation in M^2 values along with the diet-related trends of the $M_{\overline{3}}$ values predictably demonstrate that cusp relief and acuity, and the attendant functional emphases on point penetration and 'vertical' point-cutting in the ectoloph and ectolophid, are greatest by far in the most insectivorous taxa, moderate in the most folivorous taxa, and least by far, in the most frugivorous forms. The stem-feeder, *Hapalemur*, unexpectedly shows values slightly lower, rather than slightly higher, than those for the lemurid folivores. Figures 48–51 suggest that had lingual cusps been metrically compared, this situation would have been reversed. The predicted cline in values is very evident among the insectivorous-frugivorous forms, moderately evident among the frugivorous-insectivorous forms, and poorly evident among the folivorous-frugivorous and frugi-vorous-folivorous taxa.

The 'crosslophed' folivore, *Indri indri*, predictably shows greater M^2 values, and lower $M_{\overline{2}}$ values, than the 'noncrosslophed' lemurine folivorous forms.

Among the frugivore-folivores, *Lemur catta* generally demonstrates values which are relatively high. This may not be particularly surprising, in view of the fact that *Lemur catta* is a relatively very eclectic feeder, and incidentally, the only diurnal lemuriform reported to eat insects occasionally in the wild [JOLLY, 1966].

Despite the fact that *Euoticus elegantulus* demonstrates a strong primary dietary preference for gums, this taxon unexpectedly shows M^2 values which are moderately higher, and $M_{\overline{2}}$ values which are exceedingly higher, than those of other highly gumivorous or frugivorous taxa. This great similarity of *Euoticus* to the more highly insectivorous forms is demonstrated repeatedly in other molar features as well. The significance of this will be discussed later in 'Conclusions'.

Daubentonia madagascariensis predictably demonstrates exceedingly low values with respect to other highly insectivorous forms. The insect component of the diet of this taxon is mostly soft insect larvae, having physical properties closer to those of fruit.

2. Crown Relief

While figures 48–68 and the species diagnoses may convey the impression that relative crown relief varies greatly among the Strepsirhini, the histogram analyses of Indices IX (fig. 24A, 25A, 26A) and XXIX (fig. 24B, 25B, 26B) indicate that this feature is fairly constant when species are compared according to molar size (fig. 24), heritage (fig. 25) and dietary preference (fig. 26).

The $M_{\overline{2}}$ values, as in the previous analyses, tend to demonstrate stronger size and heritage-related variation than do the values for M^2. Upper molar values, on the other hand, appear to vary more with dietary preference, and in a manner similar to, but far more subdued than, the indices of cusp relief and acuity. All variation in the M^2 and the $M_{\overline{3}}$ is greatly muted, however, and demonstrates relatively poor separation according to size, heritage or diet. The lack of a greater difference in values between those taxa subsisting on relatively tough, abrasive foods (insects and leaves) and those taxa consuming relatively highly deformable and less abrasive foods (fruits, gums and larvae) is especially surprising, and would appear to indicate that crown relief as it relates to molar area is either not especially crucial to the functional longevity of molar teeth, or that this feature (for whatever morphogenetic reason) is relatively less sensitive to selection pressures to change it.

3. Crest Length, Curvature and Sharpness

a) Crest Length

Histogram analyses of Indices X (fig. 27A, 28A, 29A) and XXX (fig. 27B, 28B, 29B) indicate how molar crest development and the consequent emphasis on molar point cutting vary with molar size (fig. 27), heritage (fig. 28), and dietary preference (fig. 29). Values for the M^2, and especially the $M_{\overline{3}}$, demonstrate the strongest variation with heritage, and most notably reveal that the variably crosslophed Indriidae dominate the other taxa with regard to the total length of molar cresting. Size-related trends in both the M^2 and $M_{\overline{3}}$ values are nil or very weak, as is the diet-related variation in the $M_{\overline{3}}$ values. The values for the M^2 do show a moderate to weak pattern of diet-related variation, predictably demonstrating the highest values for the variably crosslophed indriids and for the most insectivorous taxa, moderate values for the remaining folivorous forms, lower values for the stem-feeder, *Hapalemur*, and still lower values for the most highly frugivorous taxa. Interestingly, *Phaner* and *Euoticus* show M^2 values which are very high with respect to the other frugivorous-insectivorous forms. The weak diet-related variation of the M^2 and $M_{\overline{3}}$ values (especially for the more frugivorous taxa) can probably be largely explained by the fact that measurements of crest length do not accurately reflect the sharpness or the salience of crest edges, which are parameters crucial to the efficiency of point-cutting. Figures 48–68 and the species diagnoses clearly demonstrate that the more highly frugivorous taxa possess relatively blunt, rounded and low relief crest edges. These features more strongly contrast this dietary group with the more insectivorous and folivorous taxa, yet they are difficult to measure.

b) Crest Curvature

Figures 48–68 and the species diagnoses demonstrate that occluding molar crests greatly vary interspecifically with regard to patterns of reciprocal curvature and/or differential orientation.

The indices of reciprocal crest curvature were chosen after balancing the considerations of molar occlusal relationships against those of the practicality of taking the measurements necessary for these indices. It is felt that while these indices are adequate, they are not totally satisfactory. Among other reasons, these indices are deficient because they do not reflect patterns of differential orientation of occluding crests. Careful examinations of both the morphology of the molars and the line drawings strongly suggest that patterns of reciprocal curvature and/or differential orientation of occluding molar crests vary more strongly with dietary preference than the indices would appear to indicate. Thus, if anything, these indices are very conservative indicators of the diet-related variation in these features.

Indices XIV and XXXIV reflect the degree to which the lingual crests emphasize reciprocal edge curvature in a plane perpendicular to the cervical plane of the molar, and the degree to which 'vertical' point-cutting is emphasized. The histogram analyses of Indices XIV (fig. 30, 31A, 32A) and XXXIV (fig. 30B, 31B, 32B) compare species according to molar size (fig. 30), heritage (fig. 31) and dietary preference (fig. 32), and reveal that the values for M^2 best separate out according to dietary preference, while those for $M_{\overline{3}}$ vary relatively strongly with heritage.

The diet-related variation of the M^2 values predictably demonstrates that the emphasis on 'vertical' point-cutting is greatest in the most insectivorous forms, moderate in the most folivorous taxa, and least in the most frugivorous forms. The variably crosslophed indriids predictably show values higher than those of the other frugivorous-folivorous and folivorous taxa, as does the stem feeder, *Hapalemur*.

For reasons already given, *Daubentonia* shows values much lower than those of other highly insectivorous forms, while *Euoticus* markedly differs from other frugivorous taxa in a manner similar to that of molar features previously discussed.

The relatively strong heritage-related variation in $M_{\overline{3}}$ values reveals that the galagines, lorisines, and especially indriines consistently predominate in value magnitudes. While moderately high values for the variably crosslophed indriines would be predicted, the great predominance that the indriines display in $M_{\overline{3}}$ values may be partly due to the fact that the modified molar morphology and occlusion of the Indriidae required that the 'entocristid', and not the protocristid, be measured for crest curvature. The relatively very high $M_{\bar{3}}$ values for the indriines may thus result from the fact that non-homologous molar features are being compared.

While the sum of Indices XIII and XXXIII for most species is similar (i.e., the reciprocal curvature of the lingual molar crests of most species demonstrates a similar component in a plane parallel with that of the crown cervix), the ratio of Indices (XIII + XXXIII)/(XIV + XXXIV) does demonstrate considerable interspecific variability (i.e., the component of the reciprocal curvature of the lingual molar crests in a plane perpendicular to that of the crown cervix does vary greatly between species). The higher the absolute value of this ratio, the greater is the relative emphasis on 'horizontal' point-cutting; the lower the absolute value of this ratio, the lower is the relative emphasis on 'horizontal' point-cutting.

Histogram analyses were carried out on the ratio of Indices (XIII + XXXIII)/(XIV + XXXIV) (fig. 33, 34). The ratio values appear to vary weakly to moderately with size (fig. 33), heritage (fig. 34A) and dietary preference (fig. 34B). While size and heritage influences cannot be discounted, several interesting aspects of the diet-related variation emerge. It will be noted that the most highly insectivorous taxa predictably demonstrate consistently lower values for this ratio, while there appears to be a predictable tendency for ratio values to increase greatly to the highest levels as fruit intake predominates. The more folivorous nonindriids appear to predictably emphasize moderately high ratio values, while the stem-feeder, *Hapalemur*, predictably shows somewhat lower ratio values than the lemurid folivores. *Daubentonia* predictably deviates from the pattern set by the other highly insectivorous taxa.

Despite the fact that the molars of the most frugivorous taxa demonstrate the greatest predominance of 'horizontal' crest curvature, their actual capabilities in 'horizontal' point-cutting are quite low. This is due to the fact that molar crest edges of highly frugivorous taxa are exceedingly blunt and rounded, and demonstrate a reduction in overall reciprocal curvature, which compromises both cutting efficiency and food escapement.

The indriid taxa should predictably demonstrate ratio values which are moderately lower than those for the lemurid frugivores and folivores. The indriids, however, unexpectedly demonstrate ratio values that are exceedingly low. The possible reasons for this have already been discussed.

Interestingly, the highly gumivorous taxa, *Phaner* and *Euoticus*, unexpectedly show very low ratio values with respect to other highly frugivorous forms. The significance of this fact is uncertain.

4. Basin Configurations

A perusal of figures 48–68 and of the species diagnoses reveals the great variety of configurations demonstrated by the trigon and talonid basins.

The histogram analyses an Indices XV (fig. 35A,36A,37A) and XXXV (fig. 35B, 36B, 37B) demonstrate the size-, heritage-, and diet-related variation (fig. 35, 36, 37, respectively) in the mesio-distal narrowing of the trigon and talonid basins, and primarily reflect the nature of point-cutting carried out around these basins. Low index values indicate that basins are greatly narrowed mesio-distally, and reflect an emphasis on reciprocal edge curvature and/or differential crest orientation associated with 'vertical' point-cutting. High index values indicate that basins are mesio-distally relatively wide and reflect an emphasis on reciprocal edge curvature and/or differential crest orientation associated with 'horizontal' point-cutting.

The values for the M^2 vary relatively strongly with heritage, while those for the $M_{\overline{2}}$ vary most strongly with dietary preference.

The heritage-related M^2 values show that the Lemurinae predominate in value magnitudes, followed closely by the Indriidae and Cheirogaleinae. The lowest values are shown by the Lorisidae.

The diet-related values for the $M_{\overline{2}}$ predictably demonstrate that mesio-distal basin narrowing is greatest in the most insectivorous taxa, somewhat less evident in the variably crosslophed indriids, progressively less demonstrated in the stem-feeder, *Hapalemur*, and in the most frugivorous taxa, and least evident in the most folivorous (nonindriid) taxa. *Daubentonia* and *Lemur catta* differ from other taxa in their respective dietary categories for reasons already given, while *Euoticus* may show similar tendencies to those already described.

Histogram analyses of Indices XVI (fig. 38A, 39A, 40A) and XXXVI (fig. 38B, 39B, 40B) graphically demonstrate how the mesio-distal concavity and confinement of the trigon and talonid basins vary with molar size (fig. 38), heritage (fig. 39), and dietary preference (fig. 40). These indices reflect the patterns of incusion and point-cutting, respectively, carried out in and around these basins. Within the range of index values from 40 to 100, values approaching 40 indicate that basins are mesio-distally greatly concave and highly confined, and that basin incusion surfaces are relatively greatly concave; these values thus reflect the relatively great emphases on point penetration and 'vertical' point-cutting, respectively, carried out in and around these basins. Values within the 40 to 100 range which approach 100 indicate that basins are mesio-distally only weakly concave and poorly confined to unconfined, and that basin incusion surfaces are subplanar to planar; these values thus reflect the relatively great emphases on crushing and grinding and 'horizontal' point-cutting, respectively, carried out in and around these basins. Values far above 100 indicate that 'basins' are actually mesio-distally convex and unconfined, and that 'basin' incusion surfaces are subplanar. Such high values thus reflect molars which approach the crosslophed (or bilophed) condition, and which emphasize food pleating in connection with point-cutting.

The values for both the M^2 and $M_{\overline{2}}$ appear to vary similarly and most strongly with dietary preference, and predictably demonstrate the following hierarchy: for the most highly insectivorous taxa, values generally range between 40 and 60; for the stem-feeder, *Hapalemur*, values range from 60 to 75; for the most highly folivorous (nonindriid) taxa, values fluctuate between 80 and 90; for the most frugivorous taxa, values similarly range roughly between 80 and 100; for the most highly folivorous indriine, *Indri*, values soar far above 100. These very high values for *Indri* suggest that increased bilophodonty in the indriines is related to increased folivory.

Daubentonia and *Lemur catta* differ from other taxa in their respective dietary categories for reasons previously given, while *Euoticus*, as with other molar features, shows relatively great similarities to the more insectivorous taxa.

5. Buccal Phase Facet Orientation and Molar Torsion

The M^2 of the Strepsirhini were surveyed to explore the patterns of variation in molar torsion (Index XIX) and buccal phase facet orientation (Index XX).

Despite some suspected measurement error, histogram analyses of Index XIX (fig. 41A, 42A, 43A) compare species with respect to molar size (fig. 41A), heritage (fig. 42A) and dietary preference (fig. 43A) and demonstrate that values vary strongly with molar size, but especially strongly with heritage. The heritage-related variation shows that the expression of lingually directed molar torsion is greatest in the galagines, moderate in the cheirogaleines and lorisines, less in the indriines, still less in the lemurines, and least by far in *Daubentonia*. The archaeolemurines are distinctive in emphasizing buccally directed molar torsion.

An inspection of the mandibles of the Strepsirhini indicates that torsion in the lower M2's is roughly reciprocal with that in the upper M2's. This suggests that a similar relationship tends to be maintained interspecifically between occluding features of the upper and lower molars.

The orientation of buccal phase facetting in the M^2 was relatively difficult to measure in many taxa, due largely to the poor definition of this feature. Histogram analyses of Index XX (fig. 41 B, 42 B, 43 B) are therefore probably heavily influenced by measurement error. While these histogram analyses indicate that buccal phase facet orientation fluctuates moderately in a manner only weakly related to molar size (fig. 41 B), heritage (fig. 42 B) or dietary preference (fig. 43 B), the line drawings in figures 48–68 demonstrate that this feature is actually fairly constant throughout the strepsirhines. The drawings thus suggest that occlusal relationships between features of the M^2 and $M_{\overline{2}}$ are relatively stereotyped throughout the group during buccal phase chewing.

B. Ontogenetic Patterns of Molar Wear and Molar Reorientation

Of the species samples studied in regard to ontogenetic patterns of molar wear and molar reorientation in the M^2, only that of *Lepilemur mustelinus* demonstrated strong correlations between molar wear (Index XXV) and molar torsion (Index XIX), between molar wear (Index XXV) and buccal phase facet orientation (Index XX), and between molar torsion (Index XIX) and buccal phase facet orientation (Index XX).

The regression and bivariate correlation analyses between molar wear (Index XXV) and molar torsion (Index XIX) in *Lepilemur* (fig. 45) demonstrate that these two features are strongly and positively related, i.e., as molar wear increases, molar torsion or buccal eruption also increases. Similar analyses between molar wear (Index XXV) and buccal phase facet orientation (Index XX) in *Lepilemur* (fig. 46) indicate a moderately strong but negative relationship between these two features, i.e., as molar wear increases, the horizontal component of the orientation of buccal phase facetting also increases. The final set of analyses, between molar torsion (Index XIX) and buccal phase facet orientation (Index XX) in this species (fig. 47) demonstrate that these two features are strongly and negatively related, i.e., as molar torsion or buccal eruption increases, the horizontal component of the orientation of buccal phase facetting also increases.

The significance of these patterns of relationships in *Lepilemur* will be considered in the 'Conclusions'.

VII. Conclusions

A. Diet-Related Molar Adaptations

The morphological and metrical data appear to demonstrate an often inextricable interrelationship between molar size, heritage and dietary preference as these factors relate to molar form. The web of selective forces which operate to modify molar morphology in increasingly insectivorous lineages, for example, also tend to select for small molar (and body) size. This observation, however, should not obscure the fact that within a given range of molar size, species do differ, often considerably, in molar morphology (see, for examples, the molars of *Lepilemur mustelinus*, *Cheirogaleus major* and *Arctocebus calabarensis*). Viewed differently, molar form within each family or subfamily bears an unmistakable similarity. This is perhaps most notably seen in the indriines and galagines. Within families and subfamilies too, however, molar morphology does vary, and again, often considerably, as witness the molars of *Lemur catta*, *Lemur fulvus*, *Lepilemur mustelinus* and *Hapalemur griseus* within the Lemurinae, or the molars of *Loris tardigradus* and *Perodicticus potto* within the Lorisinae. It is here suggested that the residual variation in molar morphology seen within given ranges of molar size or within given higher taxa (i.e., families or subfamilies) can be explained by dietary preference and, more specifically, primarily by the physical properties of preferred foods.

Eight pairs of indices and another (unpaired) index were utilized in testing the adequacy of my models of diet-related molar adaptation. Of these 17 indices, 7 (VI, VII, VIII, XIV, XVI, XXXV, XXXVI) strongly vary with dietary preference, 3 (XV, XXX, XXXIV) strongly vary with heritage, and one (XXVIII) strongly varies with molar size. If indices which vary moderately with 2 of the 3 factors are included, we find 10 (VI, VII, VIII, X, XIV, XVI, XXVI, XXVII, XXXV, XXXVI) which vary moderately to strongly with dietary preference, 4 indices (X, XV, XXX, XXXIV) which moderately to strongly vary with heritage, and 3 indices (XXVI, XXVII, XXVIII) which vary moderately to strongly with molar size. With respect to the 9 classes of molar features tested, values for either or both the M^2 and $M_{\overline{2}}$ vary most strongly with dietary preference in 6 classes and at least moderately with dietary preference in 8 classes. Values vary moderately to strongly with heritage in 4 classes of molar features, while moderate to strong variation related to molar size also occurs in 4 classes.

Those molar features which vary moderately to strongly with dietary preference consistently do so in a manner very similar to that predicted by my hypotheses, while features which are moderately to strongly related to heritage tend to combine in ancestral morphotypes in patterns consistent with my models of molar adaptation.

The illustrative, descriptive, and metrical data appear to indicate that the following types of diet-related molar adaptations respectively tend to characterize restrictive insect-, fruit-, leaf-, and stem-feeders.

1. In Restrictive Insect-Feeders

Cusps demonstrate very high relief and acuity, are conical, and tend to emphasize the efficiency of point penetration.

Crest orientation, moreover, emphasizes a great vertical component, while occluding crests are very long and sharp and are greatly reciprocally curved and/or differentially oriented primarily in a fashion emphasized in 'vertical' point-cutting. These features reflect an emphasis on 'vertical' point-cutting and attendant escapement patterns.

Basins are deep, mesio-distally very narrowed and confined, with reciprocally concave surfaces oriented oblique to the chewing force. They primarily facilitate the efficiency of point penetration, as well as that of 'vertical' point-cutting and attendant escapement patterns.

2. In Restrictive Fruit-Feeders

Cusps are very low, blunt, squat, and rounded, and only emphasize low penetrative functions.

The orientation of crests greatly emphasizes a horizontal component, while the length, sharpness, and overall reciprocal curvature and/or differential orientation of occluding crests are greatly reduced. These features all point to a great de-emphasis of point-cutting.

Basins tend to be rounded, very shallow and mesio-distally nearly unconfined, with weakly convex or concave surfaces oriented basically normal to the chewing force. Basin function appears to emphasize crushing and grinding.

3. In Noncrosslophed Restrictive Leaf-Feeders

Cusps display only moderate relief and acuity, and are squat and greatly elongated. These cusps bear crests in a pattern which facilitates 'horizontal' point-cutting.

Crest orientation tends to emphasize a moderate horizontal component while strongly emphasizing a mesio-distal component. Occluding crests, moreover, are moderately long and very sharp, and are greatly reciprocally curved and/or differentially oriented primarily in a fashion emphasized in 'horizontal' point-cutting. These features suggest an emphasis on 'horizontal' point-cutting and attendant escapement patterns.

Basins are mesio-distally very broad and unconfined and bucco-lingually steeply sloped, with planar surfaces oriented normal to the chewing force. They are thus capable of performing efficient crushing and grinding, while they facilitate efficient 'horizontal' point-cutting and attendant escapement patterns.

4. In Crosslophed Restrictive Leaf-Feeders

Cusps display moderate to great relief and acuity and are moderately squat. These cusps, along with lengthened crests they bear, facilitate the point-cutting of pleated food material.

In addition, crest orientation moderately to greatly emphasizes a vertical component and a component oblique to the mesio-distal axis. Occluding crests are very long and are sharp, and are greatly reciprocally curved and/or differentially oriented in a fashion intermediate between that emphasized in 'vertical' and 'horizontal' point-cutting. These features indicate that point-cutting intermediate between 'vertical' and 'horizontal' point-cutting is utilized to divide pleated food.

'Basins' in these forms are replaced by mesio-distally convex and only moderately wide transverse lophs which possess narrow subplanar surfaces oriented oblique to chewing force. These 'basins' thus primarily perform food-pleating during point-cutting, and hence facilitate the point-cutting of food.

5. In Restrictive Stem-Feeders

Cusps demonstrate moderate to great relief and moderate acuity, and are conical. These cusps appear to be capable of performing moderately efficient point penetration and accentuated bending of food material.

The orientation of crests moderately emphasizes both a vertical component and a component oblique to the mesio-distal axis while crest length is moderately reduced and crest sharpness is moderate. Occluding crests, in addition, are greatly reciprocally curved and/or differentially oriented in a fashion intermediate between that emphasized in 'vertical' and 'horizontal' point-cutting. Point-cutting intermediate between 'vertical' and 'horizontal' point-cutting and 'punching' of food appear to be functions emphasized by these features.

Basins tend to be deep, rounded, and mostly confined, with convexo-concave to reciprocally concave surfaces oriented oblique to the chewing force. Localized gaps in basin walls are emphasized. These basins primarily facilitate the point penetration, point-cutting, bending and 'punching' of food.

The following major conclusions can be drawn from this segment of the study: (1) that the diet-related molar adaptations of restrictive insect-, fruit-, leaf-, and stem-feeders can be distinguished on the basis of molar morphology alone, without invoking body mass; (2) that the physical properties of preferred foods represent significant selective influences on molar form, within the definite limits set by heritage; (3) that the molars are indeed evolutionary compromises, such that modifications in one set of molar features limit the spectrum of possible modifications in another set of molar features; (4) that considerations of food escapement are inseparable from considerations of food division; (5) that it is possible to analyze molar adaptations without resorting to body mass, and that such analyses can be profitably applied in efforts to hypothesize dietary preference from isolated fossil molar teeth.

KAY [1973, 1975] has emphasized an approach to the study of molar adaptation which involves relating the size of molar features to body mass. KAY, on the basis of his studies, has concluded that noncercopithecid primates which are highly insectivorous demonstrate functional capabilities in their molars which are virtually identical to those of highly folivorous primates. Without invoking body mass, KAY is unable to distinguish between molar adaptations relating to insectivory and those relating to folivory.

The conclusions of the present study are for the most part at variance with those of KAY, and suggest that KAY in his studies: (1) did not utilize measurements of molar function which were sufficiently diagnostic of dietary preference; (2) did not conceive of the mammalian molar as an evolutionary compromise, incapable of undergoing great modifications in one feature (or set of features) without forcing modifications in another feature (or set of features), and (3) did not sufficiently recognize the great intertaxonal variability in point-cutting and incusion performed by molar teeth.

The limitations of KAY's approach, dependent as it is on body mass, become especially apparent when applied in efforts to infer dietary preference from isolated fossil molar teeth (which provide the vast bulk of the mammalian fossil record). An approach liberated from body mass would appear to be more suitable in such situations where body mass is virtually unknown.

The consistent similarity demonstrated between the many features of *Arctocebus* and those of other highly insectivorous taxa is worth citing. The fact that *Arctocebus* specializes in feeding on caterpillars, which lack a hard cuticle or exoskeleton, while the other insectivorous forms tend to feed on insects with hard exteriors, would appear to indicate that the great functional similarities between *Arctocebus* and the other insectivorous taxa are due, not primarily to the hard exoskeletons of insects, but to the relatively great thickness and toughness of insect tissues beneath the exoskeleton.

The molar adaptations of *Indri* are also informative. Bilophodonty in *Indri* is more greatly demonstrated than in any other extant indriine, while the best available dietary data appears to indicate the *Indri* is the most folivorous taxon in its subfamily. These facts, plus an awareness of the molar modifications in the Colobinae, would appear to indicate that bilophodonty is primarily an adaptation to folivory. While molar adaptations to other dietary preferences may later be superimposed on the bilophed molar, it is suggested that the bilophed condition initially evolves primarily to increase the length of molar cresting and to provide a mechanism to pleat leafy food so as to insure that this food is efficiently point cut by molars with cusps of relatively great relief and acuity.

B. Heritage-Related Molar Features

The differential control that heritage apparently shows over the M^2 and $M_{\overline{2}}$ in several comparable features and the often stubborn persistence of many inferred ancestral molar features even in taxa which are otherwise strongly apomorphic, demonstrate that heritage, too, plays a definite part in determining molar morphology. What emerges is the observation that certain molar features are more susceptible to modification by natural selection, while other molar features are under relatively stricter genetic control (i.e., are highly controlled by heritage), and are more impervious to relatively recent selection pressures. While the genetic and morphogenetic mechanisms for this phenomenon cannot be specified, the following examples provide graphic evidence that heritage often demonstrates great control over the phenotype.

KAY, in his studies [1973, 1975] demonstrates that highly frugivorous or gumivorous noncercopithecid primates possess smaller molar teeth relative to body mass than do highly insectivorous or highly folivorous primates. Forms like *Phaner*, *Euoticus*, and *Perodicticus* bear this out very nicely. However, while the molars of *Perodicticus*, and to some extent those of *Phaner*, demonstrate commensurate morphological modifications consistent with a greater dietary emphasis on fruits and gums, those of *Euoticus*, for most features, most certainly do not! In fact, were *Euoticus* a fossil taxon, represented only by molar teeth, one would not hesitate to infer that this taxon strongly preferred insects, on the basis of its molar morphology. *Euoticus* appears to suggest that the genetics and morphogenetics of primate molars tend to render them more susceptible to evolutionary modifications in size rather than morphology.

One of the molar features which appears to demonstrate very strong variation with heritage, and which very distinctively characterizes each family (or subfamily), is molar torsion. While the significance of the variation of this feature cannot be readily fathomed, it is clear that such a feature may have some taxonomic value, especially in studies dealing with the fragmentary tooth rows of fossils.

C. Molar Features which are Intertaxonally Constant

As mentioned in the 'Discussion', crown relief (relative to molar area) appears to vary surprisingly little. Why taxa which subsist on more abrasive foods should not demonstrate far greater crown relief than those subsisting on a less abrasive diet is hard to explain. If it is true that the molar dimensions of highly frugivorous and gumivorous taxa are consistently reduced, relative to body mass, then highly frugivorous taxa do, in this sense, have less molar crown relief than either highly insectivorous or highly folivorous forms. This fact, however, is not readily reflected when relating crown relief to molar area, since these two parameters apparently vary similarly with increased frugivory.

The apparent tendencies in both the upper and lower molars to demonstrate roughly reciprocal torsion, and in buccal phase facetting to intertaxonally demonstrate a relatively constant orientation, appear to reflect the great rigidity of genetic control that exists over the occlusal relationships between molar features.

Despite the possibility that mandibular kinematics may differ interspecifically to some degree, it appears that natural selection will only tolerate variations in molar occlusal relationships within the very narrowest of ranges. Any deviations greater than this, would appear to render the intricate and delicately balanced system of occluding molar features nonfunctional.

D. Mandibular Kinematics and its Relationship to Ontogenetic Changes in Molar Wear and Molar Orientation

The intimate interrelationships between molar wear, molar occlusion and mandibular movement are graphically illustrated in the bivariate correlations of indices of these features in *Lepilemur*.

Lepilemur subsists on a diet of extremely abrasive and tough leafy material, while its molars do not demonstrate any unusual morphological features which would confer greater longevity to efficient point-cutting. Theoretically, increased molar abrasion should require an increase in the horizontal component of the orientation of buccal phase facetting so as to reorient edge honing (or facetting) so that crest edges are kept sharp: for noncrosslophed folivorous taxa, this increase should occur along a linguad axis; for insectivorous forms, this increase should theoretically occur along a progressively more mesiad axis.

With this background, the patterns of molar wear and molar reorientation in *Lepilemur* begin to assume greater significance. The molars of *Lepilemur* demonstrate unusually rapid wear due to the abrasiveness of the diet of this taxon, while the horizontal component of the orientation of buccal phase facetting increases commensurately in this form, so as to insure the ontogenetic maintenance of efficient edge honing and point-cutting. Were this process to continue for any great length of time, however, *Lepilemur* (and no doubt most other taxa in a similar situation) would be faced with a major functional problem.

The alignment of the jaw muscles and the configuration of the jaw joint apparently do not substantially change throughout the life of most mammals, yet this fact flies in the face of the apparent selective demands on *Lepilemur* to increase the horizontal component of mandibular movements during buccal phase chewing. *Lepilemur* has apparently evolved a distinctive solution to this problem, however, which satisfies the requirements of both edge honing and mandibular movement, at the same time that it meets the requirements of molar occlusion. This solution involves buccally erupting the M^2 (and presumably, lingually erupting the $M_{\overline{2}}$) throughout ontogeny, at a rate comparable to that of molar wear and consistent with the demand of edge sharpening. This adaptation insures that molar cutting edges are efficiently honed throughout ontogeny, and that occlusal relationships are maintained, while it allows the trajectory of the mandible during buccal phase chewing to remain virtually fixed throughout the ontogeny of the individual. MELLETT [1977] has described similar adaptations in creodonts.

VIII. Summary

In an effort to add to the understanding of the functional relationship between molar morphology and dietary preference, several models of molar adaptation (functionally relating molar form to the inferred physical properties of preferred foods) were developed which guided a morphological and metrical comparison of the second molars of the strepsirhine primates. The interspecific morphological and metrical comparisons provided a means of delineating the relative influences of molar size, heritage and dietary preference on molar form (see 'Conclusions'). The data corroborate the hypothesis that the physical properties of preferred food constitute very significant selective pressures on molar form, within the definite limits set by heritage.

Contrary to conclusions reached by KAY [1973, 1975], this study also indicates that the diet-related molar adaptations of restrictive insect-, fruit-, leaf-, and stem-feeders can be distinguished on the basis of molar morphology alone, without invoking body mass. Furthermore, this investigation suggests that analyses of molar adaptation, which are liberated from body mass, can be developed to study adaptations of fossil taxa for which data on body mass is lacking. That functional molar morphology represents an evolutionary compromise of both past and present adaptations and of competing and contemporaneous selection forces has also been suggested in this study. The differential control that heritage has over various molar features can also be discerned when characters are carefully delineated and compared among closely related species.

Each strepsirhine subfamily or family appears to demonstrate distinctive expressions of molar torsion, despite the fact that molar occlusal relationships between taxa appear to remain fairly constant. In addition, patterns of buccal eruption in the molars of *Lepilemur* suggest the apparent need for mammals to evolutionarily modify their molar teeth so as to comply with the functional demands of the ontogenetically and phylogenetically more conservative musculo-skeletal portion of the feeding mechanism.

IX. References

ARDRAN, G.M.; KEMP, F.H., and RIDE, W.D.L.: A radiographic analysis mastication and swallowing in the domestic rabbit, *Oryctolagus cuniculus*. Proc. zool. Soc. Lond. *130:* 257–274 (1958).

BEARDER, S.K. and DOYLE, G.A.: Ecology of bushbabies, *Galago senegalensis* and *Galago crassicaudatus*, with some notes on their behavior in the field; in MARTIN, DOYLE and WALKER Prosimian biology, pp. 109–130 (Duckworth, London 1974).

BOCK, W.J. and WAHLERT, G. VON: Adaptation and the form-function complex. Evolution, Lancaster *19:* 269–299 (1965).

BUTLER, P.M.: The milk-molars of perissodactyla with remarks on molar occlusion. Proc. zool. Soc. Lond. *121:* 777–817 (1952).

BUTLER, P.M.: Some functional aspects of molar evolution. Evolution, Lancaster *26:* 474–483 (1972).

BUTLER, P.M.: Molar wear facets of early tertiary North American primates. Symp. 4th Int. Congr. Primat., vol. 3, pp. 1–27 (Karger, Basel 1973).

CARTMILL, M.: Strepsirhine basicranial structures and the affinities of the Cheirogaleidae; in LUCKETT and SZALAY Phylogeny of the primates, pp. 313–354 (Plenum, New York 1975).

CHARLES-DOMINIQUE, P.: Eco-éthologie des prosimiens du Gabon. Revue Biol. gabon. *8:* 121–1228 (1971).

CHARLES-DOMINIQUE, P.: Ecology and feeding behavior of five sympatric lorisoids in Gabon; in MARTIN, DOYLE and WALKER Prosimian biology, pp. 131–150 (Duckworth, London 1974).

COPE, E.W.: The mechanical causes of the development of the hard parts of the Mammalia. J. Morph. *3:* 137–277 (1889).

CROMPTON, A.W.: The origin of the tribosphenic molar; in KERMACK and KERMACK Early mammals, pp. 65–87 (Academic Press, London 1971).

CROMPTON, A.W. and HIIEMAE, K.M.: How mammalian molar teeth work. Discovery *5:* 23–34 (1969).

CROMPTON, A.W. and HIIEMAE, K.M.: Molar occlusion and mandibular movements during occlusion in the American opposum, *Didelphis marsupialis*. J. Linn. Socc. (Zool.) *49:* 21–47 (1970).

CROMPTON, A.W. and SITA-LUMSDEN, A.: Functional significance of the therian molar pattern. Nature, Lond. *227:* 197–199 (1970).

DOYLE, G.A.: The behavior of the lesser bushbaby *(Galago senegalensis moholi);* in MARTIN, DOYLE, and WALKER Prosimian biology, pp. 213–232 (Duckworth, London 1974).

EVERY, R.G.: The significance of extreme mandibular movements. Lancet *ii:* 37–39 (1960).

EVERY, R.G.: Sharpness of the teeth in man and other primates. Postilla *143:* 1–30 (1970).

EVERY, R.G.: A new terminology for mammalian teeth (Christchurch, New Zealand 1972).

EVERY, R.G.: Thegosis in prosimians; in MARTIN, DOYLE and WALKER Prosimian biology, pp. 579–620 (Duckworth, London 1974).

EVERY, R.G. and KUHNE, W.G.: Bimodal wear of mammalian teeth; in KERMACK and KERMACK Early mammals, pp. 23–27 (Academic Press, New York 1971).

GINGERICH, P.D.: Molar occlusion and jaw mechanics of the Eocene primate *Adapis.* Am. J. phys. Anthrop. *36:* 359–368 (1972).

GRAHAM, S.F.: Studies on American marsupials. 1. Dental apparatus of a deme sample of *Didelphis marsupialis pigra* (Marsupialia, Didelphidae); PhD thesis, Rutgers University (1969).

GREGORY, W.K.: On the structure and relations of *Notharctus*, an American Eocene primate. Mem. Am. Mus. nat. Hist. *3:* 49–243 (1920).

GREGORY, W.K.: The origin and evolution of the human dentition (Williams & Wilkins, Baltimore 1922).

HIIEMAE, K.M. and ARDRAN, G.M.: A cineradiographic study of feeding in *Rattus norvegicus.* J. Zool., Lond. *154:* 139–154 (1968).

HIIEMAE, K.M. and CROMPTON, A.W.: A cinefluorographic study of feeding in American opossum, *Didelphis marsupialis*, in DAHLBERG Dental morphology and evolution, pp. 299–334 (University of Chicago Press, Chicago 1971).

HIIEMAE, K.M. and KAY, R.F.: Trends in the evolution of primate mastication. Nature, Lond. *240:* 486–487 (1972).

HIIEMAE, K.M. and KAY, R.F.: Evolutionary trends in the dynamics of primate mastication. Symp. 4th Int. Congr. Primat., vol. 3, pp. 28–64 (Karger, Basel 1973).

HLADICK, C.M. and CHARLES-DOMINIQUE, P.: The behavior and ecology of the sportive lemur *(Lepilemur mustelinus)* in relation to its dietary pecularities; in MARTIN, DOYLE and WALKER Prosimian biology, pp. 23–37 (Duckworth, London 1974).

JOLLY, A.: Lemur behavior (University of Chicago Press 1966).

JOLLY, A.: Lemur biology (University of Chicago Press, Chicago 1967).

KALLEN, F.C. and GANS, C.: Mastication in the little brown bat, *Myotis lucifugus.* J. Morph. *136:* 385–420 (1972).

KAY, R.F.: Mastication, molar tooth structure, and diet in primates; PhD thesis, Yale University (1973).

KAY, R.F.: The functional adaptations of primate molar teeth. Am. J. phys. Anthrop. *43:* 195–216 (1975).

KAY, R.F.: Molar structure and diet in extant Cercopithecidae (in press).

KAY, R.F. and HIIEMAE, K.M.: Mastication in *Galago crassicaudatus*, a cinefluorographic and occlusal study; in MARTIN, DOYLE and WALKER Prosimian biology, pp. 501–530 (Duckworth, London 1974a).

KAY, R.F. and HIIEMAE, K.M.: Jaw movement and tooth use in recent and fossil primates. Am. J. phys. Anthrop. *40:* 227–256 (1974b).

KRAUSE, W.J. and LEESON, C.R.: The stomach of the pangolin *(Manis pentadactyla)* with emphasis on the pyloric teeth. Acta anat. *88:* 1–10 (1974).

MARTIN, R.D. A review of the behavior and ecology of the lesser mouse lemur *(Microcebus murinus);* in MICHAEL and CROOK Comparative ecology and behaviour of primates, pp. 1–68 (Academic Press, London 1973).

MELLETT, J.S.: Paleobiology of north american Hyaenodon (Mammalia, Creodonta). Contr. verteb. Evol. *1:* 1–134 (1977).

MILLS, J.R.E.: Ideal dental occlusion in the primates. Dent. Practnr *6:* 47–61 (1955).

MILLS, J.R.E.: The functional occlusion of the teeth of insectivora. J. Linn. Soc. (Zool.) *47:* 1–25 (1966).

OSBORN, R.C.: The evolution of the occlusion of the teeth. Am. Orthod. (July, 1911).

PATTERSON, B.: Early cretaceous mammals and the evolution of Mammalian molar teeth. Fieldiana, Geol. *13:* 1–105 (1956).

PETTER, J.J.: Recherches sur l'éthologie des lémuriens malgaches. Mém. Mus.Hist. nat. *27:* 1–146 (1962a).

PETTER, J.J.: Ecologie et éthologie comparées des lémuriens malgaches. Terre Vie *109:* 394–416 (1962b).

PETTER, J.J.: The lemurs of Madagascar; in DEVORE Primate behavior: field studies of monkeys and apes, pp. 292–319 (Holt, Rinehart & Winston, London 1965).

PETTER, J.J. et HLADICK, C.M.: Observations sur le domaine vital et la densité de population de *Loris tardigradus* dans les forêts de Ceylan. Mammalia *34:* 394–409 (1970).

PETTER, J.J. et PEYRIERAS, A.: Observations éco-éthologiques sur les lémuriens malgaches du genre *Hapalemur*. Terre Vie *24:* 356–382 (1970).

PETTER, J.J. and PEYRIERAS, A.: Preliminary notes on the behavior and ecology of *Hapalemur griseus;* in TATTERSALL and SUSSMAN Lemur biology, pp. 281–286 (Plenum, New York 1975).

PETTER, J.J.; SCHILLING, A. et PARIENTE, G.: Observations éco-éthologiques sur deux lémuriens malgaches nocturnes: *Phaner furcifer* et *Microcebus coquereli*. Terre Vie *25:* 287–327 (1971).

RENSBERGER, J.M.: An occlusal model for mastication and dental wear in herbivorous mammals. J. Paleont. *47:* 515–528 (1973).

RENSBERGER, J.M.: Function in the cheek tooth evolution of some hypsodont geomyoid rodents. J. Paleont. *49:* 10–22 (1975).

RYDER, J.A.: On the mechanical genesis of tooth forms. Proc. Acad. nat. Sci. Philad. *30:* 45–80 (1878).

SIMPSON, G.G.: Studies of the earliest mammalian dentitions. Dent. Cosmos *78:* 791–800 (1936).

SUSSMAN, R.W.: Ecological distinctions in sympatric species of *Lemur;* in MARTIN, DOYLE and WALKER Prosimian biology, pp. 75–108 (Duckworth, London 1974).

SZALAY, F.S.: Mixodectidae, Microsyopidae, and the insectivore – primate transition. Bull. Am. Mus. nat. Hist. *140:* 193–330 (1969).

SZALAY, F.S.: Phylogeny of primate higher taxa; in LUCKETT and SZALAY Phylogeny of the primates, pp. 91–125 (Plenum, New York 1975).

SZALAY, F.S. and KATZ, C.C.: Phylogeny of lemurs, galagos and lorises. Folia primatol. *19:* 88–103 (1973).

VAN VALEN, L.: Deltatheridia, a new order of mammals. Bull. Am. Mus. nat. Hist. *132:* 1–126 (1966).

WALKER, P. and MURRAY, P.: An assessment of masticatory efficiency in a series of anthropoid primates with special reference to the Colobinae and Cercopithecinae; in TUTTLE Primate functional morphology and evolution, pp. 135–150 (Mouton, Hague 1975).